Geoinformatics in Health Facility Analysis

Dedicated to my parents, wife, family members, relatives and, teachers

Introduction

Health GIS is an important subdiscipline of health science and medical geography which is traditionally focused on the spatial aspects of disease ecology and health care facility analysis. With the advent of Geographic Information Systems (GIS) and remote sensing technologies (computers and sophisticated spatial analytic software programs), medical geography has been dramatically transformed like many other spatial sciences Study Area and Methodology. Keeping this in view, an attempt has to be made for the Varanasi district, India, to analyze the characteristics of population and its related variables, vis-a-vis health care facilities and the status of vector-borne diseases and malaria modeling using GIS techniques and statistical methods. The variables, i.e., literacy, sex ratio, occupational structure, etc., are selected, calculated, processed, mapped, and analyzed to find out the spatiotemporal distribution of the characteristics of population. All the data in Excel format are integrated with development block boundaries on the ARC GIS version-9.3 platform, and different choropleth outputs are produced. The existing location and number of government hospitals, i.e., PHCs, CHCs, subcenters, etc., and location of existing hospitals including private ones are shown during field survey with the help of a Global Positioning System (GPS). Result and Discussion the district as a whole, the number of medical institutions per 100,000 population has been worked out as 16.25 for the year 2009. The number of doctors/100,000 population was 5.8 in 1995–1996, which marginally increased to 5.66 in 2001, and then this figure decreased drastically and reached 3.54 in 2009. A GIS-based buffer analysis is also performed, and it is calculated that maximum villages, i.e., 73.51 % (136 villages out of 185), in the Baragaon development block fall under safe villages, whereas minimum villages, i.e., 30.82 % (only 45 villages out of 146), in the Kashi Vidyapith development block fall under the safe zone. The main aim of this study is also to determine and map the density areas of vector-borne diseases using GIS techniques. The important types of vector-borne diseases (VBD) highlighted in this study are malaria, filariasis, kala-azar, and dengue. Remote sensing data is used to identify the malaria mosquito breeding sites, such as ponds, streams, tanks, etc. The locations of kala-azar patients in different development blocks and in Varanasi are shown with the help of GPS. Through this study, the maximum cases (188) of malaria were found in Varanasi and the maximum cases of dengue (90) were recorded in city government hospitals. The Malaria Susceptibility Index (MSI) and Malaria Susceptibility Zone (MSZ) are calculated by using three statistical methods, i.e., multiple linear regression, information value, and heuristic approach. A number of thematic maps (referred to as data layers in GIS) on specific parameters which are related to the occurrence of malaria, i.e., land use, NDVI, distance to water ponds/tanks, distance to river, distance to road, distance to hospital, rainfall, temperature, and projected population density, for the year 2009 have been generated. The cumulative frequency curve of MSI values has been segmented into five classes representing near equal distribution to yield five malaria susceptibility zones (MSZ), i.e., very low, low, moderate, high, and very high, using all the three statistical methods. In the multiple linear regression method, 36.06 % of the pixel area falls under the -4000–5000 model index class and susceptibility index <-15000 contains 2.57 % of the pixel area, whereas 13.43 % of the pixel area falls under the >7000 susceptibility index class. The information value (InfoVal) analysis includes two specific steps, i.e., bivariate analysis and multivariate analysis. The area of distribution of malaria based on the information value analysis, i.e., 0.1–0.6, looks sensitive, and 29.73 % of the pixel area includes the quantities more than this amount, so this value can be defined as the crucial value for malaria, and 39.86 % and 26.29 % of the areas of malaria fall under high and very high susceptibility classes. The MSI values from the heuristic approach (weighting method) are found to lie in the range of 21–37. To compare each model for MSZ, the Qs (malaria density ratio) method is used, and it is found that the information value method having Qs=3.96 has been selected as the optimum model for malaria susceptibility zonation in the study area, whereas the Qs values for the qualitative map combination (heuristic method) and multiple linear regression method are 1.67 and 1.43, respectively. The area under a curve is used to assess the prediction accuracy qualitatively, and verification results show that in the information value case, the area under curve (AUC) is 0.696 and the prediction accuracy is 69.60 %. In the heuristic case, the AUC is 0.603 and the prediction accuracy is 60.30 %. In the multiple linear regression case, the AUC is 0.484 and the prediction accuracy is 48.40 %. Primary data is also collected from 800 respondents of 16 selected villages (two villages from each development block) in the rural part of Varanasi district to know about the utilization of health care facilities, and their results are analyzed with the help of the Statistical Package for the Social Sciences (SPSS) software. An attempt has also been made here to calculate the Hospital Requirement Index (HRI) and Hospital Requirement Zone (HRZ) through the weighting method by which it can be known as to which area needs more development in health facilities. The Hospital Requirement Index (HSI) values calculated using the weighting method are found to lie in the range of 11–23. The cumulative frequency curve of HRI values has been segmented into four classes to yield four hospital requirement zones, i.e., low, moderate, high, and very high. It is found that the areas coming under very high and high requirement classes are 46.62 % and 7.55 %, respectively, whereas 3.39 % and 42.63 % of the total areas fall under low and moderate requirement classes in Varanasi district.

Acknowledgements

It gives me immense pleasure to acknowledge the help, support, encouragement, and guidance that I have received in the course of completing this book. I would like to remember some of the important personalities for their invaluable support.

I express my deep sense of gratitude and profound indebtedness to my teacher and coauthor of this book Dr. M.S. Nathawat, Former Professor and Head, Department of Remote Sensing, Birla Institute of Technology, Mesra, Ranchi, and Professor of Geography, School of Sciences, Indira Gandhi National Open University, New Delhi, for his valuable guidance, continuous support, inspiration, and encouragement throughout my study.

I am also grateful to my co-guide and teacher Dr. S.B. Singh, Professor and Former Head, Department of Geography, Banaras Hindu University (BHU), Varanasi, for valuable suggestions whenever needed. I express my sincere thanks to Prof. K.N.P. Raju, Course Coordinator, PGDRS and GIS, and Coordinator, UGS SAP, DRS-I, and Prof. V.K. Kumra, Ex-Head, Department of Geography, and Former Registrar, Banaras Hindu University, Varanasi, for constant support and encouragement upheld my enthusiasm.

I would like to express my heartfelt thanks to Mr. Mohhamad Onagh, Former Research Scholar, Department of Geography, BHU, and Mr. M.A. Khan, District Malaria Officer, Varanasi, for their suggestions and cooperation. I am also grateful to Dr. Kshitij Mohan, Technical Assistant, Department of Geography, Banaras Hindu University, for his friendly guidance during the preparation of this book. I also express my sincere thanks to Mr. Amit Srivastava, S.T.A., Department of Geography, Banaras Hindu University, for giving me moral support.

It is hard to express my feeling of gratitude to my parents, my loving wife, my in-laws, my sisters, my grandmother, my late grandfather, and all my family members and other relatives for their eternal support and encouragement to make everything possible.

I cannot forget their immense help and support in every phase of my work. Their smiling and jolly natures have always radiated positive energy, which has constantly been my source of inspiration. Last but not the least, I express my deep sense of gratitude to all my respondents in the selected villages of Varanasi district.

Varanasi, India Praveen Kumar Rai

Contents

Author's Details

Dr. Praveen Kumar Rai, M.Sc., M.Tech., Ph.D.
Assistant Professor (PGDRS & GIS)
Department of Geography
Institute of Science
Banaras Hindu University
Varanasi, Uttar Pradesh, India

Dr. Mahendra Singh Nathawat, M.A., M.Phil, Ph.D., PGDRS
Professor of Geography and Director
School of Sciences (SoS)
Raman Bhawan, Block D
Academic Complex
Indira Gandhi National Open University
New Delhi, India

Abbreviation

AUC	Area Under Curve
CHC	Community Health Center
CPCB	Central Pollution Control Board
GIS	Geographic Information System
GPS	Global Positioning System
IRHMS	Integrated Health Management System
IRS	Indian Remote Sensing
HRI	Hospital Requirement Index
HRZ	Hospital Requirement Zone
LISS	Linear Imaging Self Scanner
LU/LC	Land Use/Land Cover
MCH	Mother Child Health
MSI	Malaria Susceptibility Index
MSZ	Malaria Susceptibility Zone
MSS	Multi-Spectral Scanner
NHS	National Health Services
NDVI	Normalized Difference Vegetation Index
PCA	Population Census Abstract
PHC	Primary Health Center
Qs	Quality Score
RS	Remote Sensing
TM	Thematic Mapper
WHO	World Health Organization
WLC	Weighted Linear Combination
VBD	Vector-Borne Disease

Chapter 1
Health Care System and Geospatial Technology: A Conceptual Framework of the Study

Abstract In this chapter an introduction to the main themes of the book, and the role health geography and geospatial technologies like geographic information system techniques, remote sensing etc. in the health sector and different aspects related with health GIS, medical geography, spatial epidemiology, concept of primary health care are very clearly discussed. The present study has been designed to analyses the pattern of health care delivery system in rural areas of Varanasi district and status of vector born disease (VBD) especially of malaria along with effects of important factors responsible for spreading the malaria area including population density of each development blocks.

Keywords Health GIS • Remote sensing • Medical geography • Spatial epidemiology • Primary health care • Vector Born Disease (VBD)

1.1 Introduction

Uncontrolled and rapid growth of population has exerted greater pressure on urban and rural amenities. It has caused various diseases and other health related problems. With the rapid increase in population as well as rising standard of living, pressure on health care facilities has also increased and has also caused an unorganized system of health care delivery facilities. Public health and health care are important concerns for developing countries and access to health care is a significant factor that contributes to a healthy population (Black et al. 2007). Most of the health care facilities are concentrated in cities area whereas in rural areas a large gap in its distribution still occurs. In rural areas, villagers have to travel 5–10 km for availing primary health care facilities (Verhasselt 1993). The present study has been designed to analyse the pattern of health care delivery system in rural areas of Varanasi district and status of vector born disease especially of malaria along with effects of important factors responsible for spreading the malaria area including population density of each development blocks. The present study also depicts a malaria model with the help of statistical methods and GIS techniques. For malaria disease area estimation not only rural area but also Varanasi city was also included. In this context it is necessary to discuss the conceptual framework for geographical analysis of

P.K. Rai, M.S. Nathawat, *Geoinformatics in Health Facility Analysis*,
DOI 10.1007/978-3-319-44624-0_1

utilization pattern of health care facilities and status of vector born disease in city area and also in different development blocks of Varanasi district.

Basically, health aspect has a multidimensional approach. Several definitions of health have been given so far. However, the WHO (1948) has tried to make a constructive meaning of health as "a state of complete physical, mental and social well-being and not only, the absence of disease of infirmity". Health can't be never static; it is dynamic and varies within a range. Therefore, it is defined as a 'flexible state of body and mind which may be described in terms of a range within which a person may sway from the condition wherein he is at the peak of enjoyment of physical, mental and emotional experience, having regards to environment, age sex and other biological characteristics due to the operation of internal and external stimuli and can recover that position without outside support' (Goel and Rajneesh 2007).

Health is influenced by a number of factors such as adequate food, housing, basic sanitation, healthy lifestyle, protection against environmental hazards and communicable diseases, the frontier of health extend beyond the narrow limits of medical care (Senthil et al. 2004).

The Planning Commission of India has described it as "Health is a positive state of well-being in which harmonious development of physical and mental capacities of the individual lead to environment of rich and full life". Perfect health is an abstraction which may not be attainable but is essential for an individual or a group or a community's strivings. Optimum health is highest level means striving for preservation and improvement of health.

A positive mental health state indicates that the individual enjoys his routine; there are no undue conflicts, not frequent bouts of depression or elevation of mood, he has harmonious relations within the family and community spheres and is not unduly aggressive. Negative health needs scientific effort for prevention and cure of diseases. The important factors for cultivation of health care are: (a) environment conductive for health full living, (b) balance diet, (c) adequate physical activity and rest as per individual needs, (d) promotive, preventive, therapeutic and welfare services, (e) suitable occupation with job satisfaction and (f) proper use of leisure and wholesome mental attitude to life.

1.2 Significance of the Study

The experiences of the use of GIS in the health sector have well been defined in health geography where epidemiological studies seek a visual representation of medical data, particularly disease data, morbidity and treatment records and also analysis of health care facilities (Escobar et al. 1997). Health atlases are an example of this early approach of the use of mapping methods in different health applications. Obviously, health of people has a close relation with the economic power of a nation. On the other hand, for the efficiency of industry and agriculture, the health of the worker is of paramount significance. The problem of the health care facilities and their utilization of any area are complex and their solution depends on proper assessment of spatial distribution of health care facilities in an unit. In our country, limited study has been performed in the above-cited aspects. So keeping this in view, an attempt has been made to analyse the health care facilities and disease problems using GIS techniques in Varanasi district, India. It may be stated that this newish approach in health sector would set up certain facts that would prove fruitful to combat the diseases.

1.3 Role of Geospatial Technology in Health Care Analysis

One of the great milestones in the development of the association between medical geography came during the 1980s, when numerous new geospatial technologies including geographic information systems (GIS), global positioning systems (GPS), and remote sensing (RS) were introduced into the fields. These technologies, combined with the increased availability and power of computing systems and software that deal with spatially explicit data, led to an explosion of research and a renewed interest in the capabilities of these two fields (Glass 2000).

GIS provides a holistic approach to public health that promotes the well-being of human populations through organizing data about who we are, where we live, and how we live within a geographic framework (Cromley and McLafferty 2002).

The area of GIS and health care has risen to prominence in the past 5–10 years with the recognition that health surveillance practices and health service allocations need to become more sensitive to the needs of people in local geographic areas (Mahmoud et al. 2014). The collection, storage and manipulation of geographic information have undergone a revolution in recent years with the development and wide spread availability of GIS software's (Mahmoud et al. 2014). Today, many health care planners and officers can benefit from education and training in the GIS field and this will give them the chance of influencing the progress of health surveillance, environmental health assessment and the geographic allocation of health resources (Curtis and Taket 1989; Murad 2004).

The call for streamlining of public health care operations as a means of achieving greater cost-efficiency is a positive sign in the development of a society. Community leaders searching for innovative methods for health care management have begun to recognize the power of a GIS in various management activities ranging from determining intervention strategies to formulating health care reforms (Albert et al. 2000). Through the geo-coding process, a GIS allows personal health data to be examined spatially so that patterns can be discerned. Furthermore, geo-referencing of personal health data can greatly enhance decisions made by public health officials (Ali et al. 2004). The tremendous potential of a GIS to benefit the health care industry is just now beginning to be realized. Both public and private sectors (including public health department, public health policy and research organizations, hospitals, medical centers, and health insurance organizations) are beginning to harness the data integration and spatial visualization power of a GIS (Nicol 1991).

GIS integrates data that designate population characteristics, socioeconomic conditions, and the landscape and analysis the spatial relationship of these factors. In addition to integrating and investigating health related data, this technology promotes data sharing through the use of standard formats and a highly efficient communication tool—the health map (DevenhaLL 2002). The GIS and the health community in India are living in two separate watertight compartments. Both need each other. The map sector of the country will find a large market for its products and services in the Health sector. The health sector will make a quantum leap in its service delivery using maps for better planning and decision making. But both are not able to talk to each other due to rigid institutional framework. The limited and sometimes non-existent commercial orientation of the government organizations (i.e. Survey of India, Indian Council of Medical Research etc.) in the country also led them to ignore the opportunities offered by new technologies like GIS.

It could be enumerated that GIS is in infancy in health sector in India as only few southern states have started mapping health indicators over physical features and infrastructure. There are some fragmented works on mapping diseases (Srivastava and Nagpal 2000). GIS takes account of spatial variation in any attributes and here it can be health investment as well as outcomes. Through the integrated data one can go for any kind of query using all sorts of data related to physical environment including topographic features, climate, vegetation, land use, water quality; settlement, various infrastructure, amenities and utilities, demographic including literacy and levels of education, socio-cultural, economic scenario including employment, and health (Park 1995).

Apart from GIS techniques, remote sensing is also very good cost-effective source to get the temporal information at regular interval about the phenomena on the ground without any physical contact (Higgs 2009). In medical geography, when such problems as the identification of endemic areas of disease, the estimation of populations at risk, and the assessment of environmental information in areas that lack baseline data or cannot be accessed arise, the use of remote sensing, especially in conjunction with other technologies such as GIS, provides an efficient and effective method of data capture. For example, satellites such as Landsat's MultiSpectral Scanner (MSS) and Thematic Mapper (TM), Indian remote sensing series satellites etc. can provide information about vegetation cover, landscape, structure and water bodies in almost any region of the globe—information that can be extremely valuable in health research that examines environmental factors in disease dissemination (Beck et al. 2000). Advantages of Remote Sensing and GIS techniques lie in the fact that health informatics could be renewed from time to time, with little pain.

GPS today is used in a variety of geospatial applications, from innovative cartography to on-board consumer automobile navigation systems (Pellerin 2006). GPS technology works through the synchronized timing of signals sent through a known array of satellites regularly in orbit above the earth's atmosphere (Pellerin 2006). These satellites transmit data that include the time the transmission was sent, the orbital path position, or ephemeris, and the rough orbital position of all GPS satellites, or the almanac. The GPS receiver, on the other hand, uses the arrival time of each data signal to calculate the distance to its respective satellite source. By using signals and calculating the distance from several satellites, each receiver can define its own position in space and time using geometric and trigonometric algorithms in a process well-known as trilateration (Daly 1993). The outcomes of trilateration are transformed to into user-friendly formats such as latitude and longitude or the location on a map. The applications of GPS have grown from military uses, such as target tracking, missile guidance, search and rescue, and investigation, to an array of civilian uses that need absolute location, relative movement, or time transfer (Pellerin 2006). One of the practical civilian uses of GPS that has gained significant attention within the last few years has been the use of GPS in health care studies (Dwolatzky et al. 2006; Joyce 2009). Although the assessment of GPS in health study has only been comparatively recently identified its use and application continues to develop (Cano et al. 2007).

1.4 Medical Geography

Medical geography is an important active sub discipline of geography which has traditionally focused on the spatial aspects of disease ecology and health care delivery. In a conventional way, medical geographers collect and analyse the data using on-the ground observations (e.g., of malarial mosquito habitats) and draw maps (e.g., of hospital catchments areas) by the hands. However, with the advent of geographic information systems (GIS) and remote sensing technologies with deal with the sophisticated spatial analytic software programs, medical geography has largely been transformed. It is now possible, for example, to make many measurements from far above the earth's surface and produce many maps of disease and health phenomena in a relatively short time (Donald et al. 2005). According to Rosicky (1965) "Medical Geography is a scientific discipline joining medicine with geography". Therefore, systematic study of the relationship among diseases attacking human beings living in their localities is the basic object of medical geography.

There are a number of geographical factors, which effect and sometimes even regulate the health and reproductive capacity of living beings, including man. The natural environment the land, the water, the air and the flora and fauna makes certain places more suitable for human habitation and healthy living than others. The type and quality of food available for human consumption are basically dependent to a greater extent on the physical and economic geographical conditions and characteristics. Poor food supply lacking in nutrients means poor health. The culture of the people is a reply of man to natural environment and is the utmost significant factor of health, cobmes within the purview of human geography. An organized study of the spatial distribution of diseases, health and ill health and their causes therefore fall within the medical geography. The concept of medical geography contains of geographical relationship in three sub-sections- diseases, nutrition and medical care (Pacione 2014).

Hunter (1974) has accurately said that medical geography is an interdisciplinary field of study related with areal disparities of disease incidence as expressed by mortality or morbidity indices and with the demonstration of possible cause-effect relationship with elements of physical, biological and socio-cultural environments in space. Distributional studies, mortality and public health were the significant search of earlier medical geography (Meade et al. 1988). There seems to be three major interrelated components in medical geography: diseases, nutrition and medical care. The main aim of studying medical care in any spatial unit develop a geographical perspective is to gain knowledge that will prevent, reduce, alleviate or cure diseases. Therefore, the geography of the medical care in spatio-temporal context is conceptually an integral and indispensable part of the medical geography. The analysis of the human environmental relationship of disease, nutrition, and medical care system in order to elucidate its interrelationship in space are a significant aspects of the medical geography. It is one of the most important goals which medical geographers are facing towards the clarification of concept and definition of the sub-disciplines.

1.4.1 Determinant of Health

Health aspect is interlinked with several influences. The influences which affect health and well-being are known as determinants of health. The important determinants are: (i) genetic configuration, (ii) level of development, (iii) life style, (iv) environment and (v) health infrastructure. All these determinants play a significant role in strenthing the health status of the people. For example with improved literacy and better employment opportunities now available, many of the health problems confronting the less developed countries have been erased.

The physical, social and biological environments of man are very important determinant of health. Poor environment sanitation, inadequate safe drinking water, excessive level of air pollution etc. are important determinant in the physical environment affecting health. Further, available and acceptable health facilities have a direct bearing on health status. If primary health care facilities are available in the vicinity of settlements and such facilities are utilized by the population, the health of individuals and communities is certainly be improved.

1.4.2 Public Health

Public health has been well-defined as "the science and art of preventing disease, prolonging life, and promoting health through the systematized efforts of society." This description was arrived at in the review established to consider the future development of the public health function including the control of communicable disease in England (Acheson 1988). The analysis was set up following failures in the system to protect the health of the public from two major outbreaks of communicable disease caused by Salmonella and Legionnaires' disease. Since then, a number of health scares have highlighted the need for continuing improvements in public health protection systems. For visualizing conceptual level of public health its definition put forward by Winslow may be quoted here. Public health may be defined as the science and art of (i) preventing disease, (ii) prolonging life and (iii) promoting health and efficiency through systematic community efforts for: (a) the sanitation of the environment, (b) the control of communicable diseases, (c) the education of the individual in personal hygiene, (d) proper organisation of medical and nursing services for the early diagnosis and timely treatment of diseases, and (e) the growth for the maintenance of health (Goel 1984).

1.4.3 Community Health

The word 'community health' aims at community diagnosis and treatment for cultivation of physical, mental and social well-being in place of traditional individual diagnosis and treatment of an individual's illness. Otherwise the term community health, a new perception, swaps the previous terms hygiene and public health. (Senthil et al. 2004).

1.4.4 Health Care

This implies not only the care of the acutely and chronically ill but also rehabilitation case findings, health maintenance, prevention of diseases and disability and health education. Thus health care emphasises a comprehensive set up which needs to be differentiated from medical care relates to care of the diseased (Datta and Kale 1969). Medical care covers a wide range of preventive, curative and rehabilitation services which are delivered by a network of individual and institutions in both private and public sittings. The pressing problems of medical care contain the maintenance of quality of care and quality of medical personal (Verhasselt 1993).

1.4.5 Disease

Disease is easier to appreciate and less abstract than health. Whereas health denotes a perfect harmony of the different body system, disease denotes an aberration of this harmony. This aberration may range from a biochemical disturbance to severe disability or death. Even a psychological dysfunction may be classified as disease. It is observed as disturbance between external and internal environments leading to departure from normal state of mental and physical health and well-being. Ecologically the factors namely, agent, host and environment are important in the study of disease pattern and there is close interrelationship among them.

1.4.6 Spatial Epidemiology

There is increasing epidemiology research evidence to the state that there is a great correlation between health and numerous features of human settlement viz. housing, work environment, residential densities, overcrowding, transportation, quality of life etc. (Nipada 2005; Bhunia et al. 2012 and 2013). The study of distribution and determinant of disease prevalence of man is called as epidemiology. Epidemiology of infectious diseases is concerned with factors that affect the probability contact between an infectious agent and a susaptible host. The descriptive epidemiology deals with the study of distribution of disease in human population while analytical epidemiology is a search for the primary causes of disease prevalence whereas Rosicky (1965) designated it as follows:

- To investigates the history of health of population and the rise and fall of diseases
- Variations in their character,
- To identify the health of the community,
- To study the working of health services,
- To estimate individual risks and chances,
- To evaluate clinical picture of chronic diseases and to describe it natural history,
- To detect syndromes and
- To examine causes of health and diseases.

Spatial epidemiology studies are concerned with finding good description of spatial incidence of diseases as well as the modelling of such incidence. One way of describing the spatial distribution of a certain disease is by visualizing the GIS choropleth maps that show the spatial distribution of such disease. In such maps, disease rates are plotted over the base map to define the areas that are highly affected from the related disease. Further analysis and modelling of the spatial incidence of diseases can be carried out using for example density estimation technique that is used to predicting the spatial variation in diseases risk. (Hoogh et al. 2001; Mahmoud et al. 2014).

No doubt, GIS is measured as a valuable tool for responding the preceding questions. For example, GIS can express the actual location of health events, and then overlay analysis can be used to make new spatial relationships and to tag the various socio-economic and environment information to the health data (Meade and Earickson 2000).

1.4.7 Concept of Primary Health Care

As a part of Alma Ata declaration, World Health Organization (WHO) (1978) has declared that 'Health is a fundamental human right and that are attainment of the highest possible level of health, is a most important worldwide social goal, whose realization requires the action of many other social and economic segments in addition to health sectors'.

In order to explain the concept of primary health care more clearly, paragraph VI and VII of the declaration of Alma Ata, perceives it as vital health care based on practical scientifically sound and socially acceptable method and technology made universally accessible to individual and families in the community through full community participation and at a cost that the community and country can afford to sustain (WHO 1978). It is the level of contact of individuals, the family and the community with the national health system, constituting the first elements of a continuing health care process. It addresses main health problem caused in the community providing primitive, preventive, curative and rehabilitative services accordingly. It includes the following aspects in details:

- Education concerning preventing health problem and methods of preventing and controlling them.
- Promotion of food supply and proper nutrition,
- An adequate supply of safe water and basic sanitation,
- Maternity and child health including family planning,
- Immunization against the major infectious diseases,
- Proper treatment of common diseases and injuries, and
- Prevention and control of locally endemic diseases, and provision of essential drugs.

Srinivasan (1984) defined primary health care "as comprising of basic curative, care, simple diagnosis and treatment, referral of complicated cases to a higher level, preventive care and vital educational measures". He has also observed primary health care as ambulatory health care that may take place at either the pre-professional level or the first contact professional level. The pre-professional stage includes self-care and lay care of family and neighbors. The first contact professional stage may comprise contact within the health or "mental health" system.

Logically primary health care is perceived as preventive endeavor. Basically, primary medical care has three roles, first management of their ills by replying to the perceived needs for medical care. Secondly, providing continuing care chronic diseases and helping them to adopt to their durability's and thirdly, that of providing preventive and educational services.

A combined report of WHO and UNICEF (1978) has also tried to define the concept of primary health care. It is designated as "primary health' care delivers primitive, preventive, curative and rehabilitative services". These services differ from country to country and community to community as it is interconnected with the socio-economic development of the country or the community concerned". Primary health care comprises following services as rendered by the government:

- Maternity and child including family planning,
- Promotion of suitable nutrition and sufficient supply of safe water,
- Basic sanitation,
- Immunization against main communicable diseases,

- Prevention and control of locally endemic diseases,
- Educational regarding the prevailing health problems and the methods of preventing and control them, and
- Proper treatments for common diseases and injuries.

According to Mohler (1978), the primary health care rests on the following fundamental thoughts:

- Important role in health planning, operation and control of primary health care must be taken by the community,
- Primary health care will succeed if it is combined at all levels as a part of the national development plan,
- Health care has to be fairly spread,
- National resources are desired to support primary health care, and
- The presence of supportive health services information is essential for providing primary health care.

Noack (1980) has specified that primary health care has no internationally accepted definition related to loose series of item, including primary medical care, which defines the action of doctors at the periphery and basic health services which promotes an interweaving services of health action among the people outside hospitals. "It is now fully realised" that the best way to distribute health care to the vast majority of underserved rural people and urban poor is to develop active primary health care services, maintained by an appropriate referred system. Primary health care has, therefore, become the core of the strategy for attaining the objective of health services of any area (Rohde 1990).

In extending the coverage of primary health care, the position of using traditional system of medicine has become evident. Practitioners of traditional system of medicine represent an enormous source of health man power, especially in the rural parts of the developing countries. These practitioners may be used with benefit for the distribution of primary health care services after suitable training and re-orientation (Ghosh and Mukharjee 1989).

The health care establishes the management sector and includes organizational matters. In India health care system is considered by the following major sectors or agencies which differ from other by the health technology applied and by the source of funds operation (Senthil et al. 2004).

These are:

1.4.7.1 Public Sectors Health Units

- Primary health care, primary health centres, sub-centres.
- Hospitals/health centres, community health centres, specialist hospitals, teaching hospitals.
- Health insurance scheme, employee/staff state insurance, central government health scheme and programmes.
- Other agencies-defense services, railways etc.

1.4.7.2 Private Sectors Health Units

- Private hospitals, nursing homes and policlinics.
- General practitioners and clinics.
- Indigenous system of medicine-ayurveda and siddha, unani and homeopathy.
- Voluntary health agencies.
- National health programme.

In 1977, the Government of India started a rural health system based on the principle of placing people's health in people hand. It is three system of health care delivery in rural region based on the recommendations at international conference at Alma-Ata in 1978 to set the goal of an acceptable level of health for all the people of the world by the year 2000 through primary health care approach. Keeping this in view the WHO goal of health for all by 2000 A.D., the Government of India developed a National Health Policy based on primary health care approach. It was approved by the parliament in 1983. The National Health Policy has laid down a plan of action for shaping the present rural health infrastructure with specific objective to be achieved by 1985, 1990, 1995 within the framework of the Sixth (1980–1985) and Seventh (1985–1990) Five Year Plan and 20 point programme.

1.4.7.3 Village Level

One of the basic views of primary health care is universal coverage and equitable distribution of health resources. It obviously emphasized that health care must penetrate into the farthest reaches of rural areas and that everyone should access to it. To implement this plan at the village level. The following schemes are in operation:

- Village health guides scheme;
- Training of local dais and workers.

1.4.7.4 Sub-Centre Level

The sub-centre is the peripheral outpost of the existing health care delivery system in rural area. They are being well-known on the basis of one sub-centre for every 5000 population in plain area and one for every 3000 population in mountainous or hilly tribal and backward region.

1.4.7.5 Primary Health Centre Level

The Bhore Committee in 1946 presented the concept of primary health centre as a basic health unit to deliver as close to the people as possible on integrated curative and preventive health care of the people. The National Health Plan (1988) suggested organisation of primary health centre on the basis of one PHC/for every 30,000 rural population in the plain regions and one PHC for every 20,000 population in mountainous or hilly and backward areas for more effective coverage.

1.4.7.6 Community Health Centre Level

Each community health centre covering a population of one lakh (one in each development block) with 30 beds and specialists in surgery, medicine, obstetrics and gynecology and pediatrics with X-ray and laboratory facilities established for strenthing, preventive and primitive aspects of health care.

During the 65 years of Independence major national resources through nine consecutive development plans have been invested in the creation of an infrastructure intended to provide elementary primary and referral health services throughout the country. The country had to achieve 100 % of the target set for establishment of sub-centres and PHC's by the end of the Seventh Plan period. However, because of the resource constraint only 50 % of the community health centres could be recognized by the year 1990. Coordinated efforts were made under various rural health programmes to deliver effective and efficient facilities to the rural populations. By 1990, the country is aided with a massive network of some 120,000 rural sub-centres accomplished by one or two paramedical workers at each centres with one or two medical officers, 3500 community health centres with specialty services and 30 inpatient beds.

Development of this enormous infrastructure along with training and equipping of man power has been a major investment and commitment to public health system during five decades of India's development programme. Till today curative services are only used by 20 % of less of the people for most common illness as mentioned in studies conducted by ICMR (Annual Report 1987), Operations Research Group (Khan 1980) and Indian Market Research Burva (Rohde 1990). Use of antenatal care, delivery services, prophylaxis for Anemia and Vitamin-A deficiency were also uniformly low. Only family planning services were followed seriously with relentless energy through annual objectives and multi-sectoral provision in the employment of acceptors (Annual Report 1980). However overall the public health infrastructure has not yet functioned to distribute a wide range of health services to a majority of people.

According to Flabault (1972), primary health care implies basic health care interpreting to the necessity of the family in the home and village environment and for its simple every day wants in the medical and health care. It is an effort to provide an integrated basic health delivery system to rural masses. To make primary health care more operative and acceptable, following subjects must be given genuine attention:

- Health assessment (individual and community),
- A suitable range of health activities,
- Continuing of health care services,

- Coordination of health care system,
- Progressive health care services and
- Family orientation in health care system.

According to Flabault (1972), the vital elements of suitable primary health care are as follow:

- Commodities: which should be safe, acceptable, achievable and economical,
- Primary health worker : which should be in acceptable number, specifically trained and functionally capable,
- Information system should be acceptable,
- Proper health support system.

1.4.8 Five-Year Plans-National Health Policy in India

The health policy executed earlier by the government could be designated in terms of the priorities fixed for various aspects of health and the objectives which it required to attain. Realising the health as an important influential factor in the utilization of manpower and in the upliftment of the economic conditions of the country considerable stress has been given upon the health programmes in the Fifth Year Plans.

1.4.8.1 First Five-Year Plan (1951–1956)

At the time of preparation of the First Five Year Plan, the health standard of the people was most unpleasant and distressing and many were sufferers of communicable and other dreaded. The death rate was very high in such a bewildering background, the First Five Year Plan estimated at improving the health standard of the population through various programmes the details of which are given below:

- Improvement of water supply and sanitation system,
- Control of malaria disease,
- Preventive health care facility of the village population through health centre,
- Health facilities for mothers and children,
- Health education and training to everybody especially in rural area,
- Self-adequacy in drug and equipment and
- Family planning and population control.

1.4.8.2 The Second Five Year Plan (1956–1961)

The Second Five Year Plan was an extension of the development effort started in the First Plan. The professed aims of health programmes during the second Plan was to extend the existing health services, to bring them within the reach of all the people and to encourage a progressive development in the national health. The detailed objectives were as follows:

- Provision of sufficient institutional facilities to serve as a base for organizing health services in any area.
- Development of technical manpower in the health sector,
- Prevention from the communicable diseases,
- Improvement of environment sanitation and
- Awareness of family planning.

1.4.8.3 The Third Five Year Plan (1961–1966)

The Third Five Year Plan was formulated with the broad objective of expanding health services for securing a progressive improvement in the health of the people and for creating conditions favorable to greater efficiency and production in continuation of the programmes originated during the earlier plan periods. The Third Plan laid greater importance on preventive public health services and eradication and control of communicable disease. The comprehensive objectives of the Third Five Year Plan regard to health were cited below:

- Control of communicable disease,
- Development of environmental sanitation in rural and urban area
- Improving facilities for training of medical and health employees,
- Provision of adequate medical institutional facilities,
- Family welfare and maternity and child health services and
- Health education and nutrition awareness.

1.4.8.4 The Fourth Five Year Plan (1969–1974)

During the Fourth Plan, high importance has been given to the control of malaria, tuberculosis, leprosy, trachoma, eradication of smallpox, strengthening of Primary Health Centres (PHCs) and family planning. The key objectives of the Fourth Plan were:

- Establishment of PHC's and sub divisional and district hospitals,
- The integration and execution of programme relating to control of communicable diseases such as tuberculosis, malaria, leprosy and small pox and
- Training of health functionaries and success of self-sufficiency in the manufacture of drugs and equipment's.

1.4.8.5 The Fifth Five Year Plan (1974–1979)

The Fifth Five Year Plan laid much importance on the improvement of health care services in the rural areas and mostly it considered the provision of health facilities as a main component of the minimum need programme. The objectives of the Fifth Plan in regards to health were as follows:

- Increasing the network of medical facilities and health services in the country,
- Increasing availability of health services and medical facilities in rural areas,
- Correction of inequalities in health services between urban and rural areas,
- Strengthening of national programmes for eradication of communicable diseases especially malaria and small pox,
- Laying greater stress on provision of sufficient supply of drinkable water and the disposal of solid wastes,
- Improving the quality of health services by securing the availability of trained medical and paramedical personnel at each PHC level and,
- Improving the quality of and providing the necessary rural orientation to the medical and para-medical personnel.

1.4.8.6 Sixth Five Year Plan (1980–1985)

Under the sixth Five Year Plan, the minimum need programmes were sought to be intensified with a more concerned research for developing alternative models of health care on the basis of Srivastava Committee Report (1978). The following were envisioned as the principles guiding the preparation of the health components in the Sixth Year Plan:

- Better health care-services must be provided for the poor people,
- A community based programme of health care and medical services should be launched on a priority basis, and
- No linear expansion of curative specialization services in urban areas was to be generally allowable.

1.4.8.7 The Seventh Five Year Plan (1985–1990)

The Seventh Five Year Plan placed much importance on improving the health status of the population including standard of living and development of human resources. The government considered the following points for development of health care facilities as well as human resource:

- The facility of primary health care facilities particularly in rural areas,
- Drug policy,

- Eradication programme for major communicable and non-communicable diseases,
- Facilities for medical, para-medical education and training, and
- The delivery of family planning, MCH immunization and related services.

1.4.8.8 The Eight Five Year Plan

It is towards human development that health and population control are noted as two of the six important objectives of this plan. Health facilities must reach the whole population by the end of the eight plans. The '*Health For All*' (HFA) paradigm must take into account not only high risk weak groups i.e. mothers and children's but must also emphasis sharply on the under honored segments. This can only be attained through emphasizing the community based health system reflected in our planning with about 30,000 populations as a standard for primary health care.

1.4.8.9 Rural Health Programme

Development and establishment of rural health infrastructure through a three tier systems of sub-centres, primary health centres (PHC's) and community health centres (CHC's) for distribution of health and family welfare services to the rural community was continued during the Seventh Plan. But lack of hospital building, shortage of manpower and sufficient provision of drug supplies and equipment generate major impediments to full operationalisation of these units.

The major approach and strategies for development of rural health during the Eight Plans were:

(a) Consolidation and operationlisation rather than major expansion of the network of sub-centres, PHC's and CHC's so that their recital is optimized. This would be accomplished through:

   - Strengthening of physical facilities including achievement of building of the centres and staff quarters,
   - Facility of related essential equipment as per standard list,
   - Recruitment of employees within a defined time frame and in services training of staff,
   - Ensuring supply of essential drugs, dressing and other materials, and
   - To monitor the progress of execution of health programmes at the regional and local levels, a health information management system is developed and used.

(b) The target regarding setting up of sub-centre, PHC and CHC on the basis of population norm are indicative only. The state governments will be given flexibility in creating these units as per the local need depending on geographical and population considerations, resources, manpower availability etc. In opening new centres, the needs of tribal population and communities living in tough and inaccessible terrains will be given first priority.

(c) The rural hospitals and dispensaries suitability merged into sub-centres, PHC, CHC as the case may be thereby assimilating them into primary health care system.

(d) The backlog at sub-centres, PHC's, and CHC in many states is staggering and the resources obligatory to meet the target are difficult in near future. In view of this the entire policy of establishment of sub-centre, PHC and CHC with the present norm should be studied and new policy options should be developed to make the primary health care available, acceptable and affordable to all.

1.4.8.10 The Ninth Five Year Plan (1995–2000)

The objectives of the Ninth Plan period comprised:

- Meeting all the felt-needs for contraception, and
- Reducing the child and mother illness and death so that there is a reduction in the desired level of fertility.

The policies during the Ninth Plan were:

- To evaluate the needs for reproductive and child health at PHC level and undertake area specific micro-planning, and
- To deliver need-based, client-centred, demand-driven high quality, integrated reproductive and child health care.

The programs focused towards:

- Providing additional assistance to poor performing districts recognized on the basis of 1991 census,
- Ensuring continuous supply of important drugs, vaccines, contraceptives of suitable quality and quantity, and
- Promoting male contribution in the planned parenthood movement and increasing of acceptance of vasectomy.

1.4.8.11 The Tenth Five Year Plan (2000–2005)

The areas of care in the Tenth Five Year Plan encompassed the reorganization and restructuring of existing health care infrastructure, including infrastructure for delivering health services at primary, secondary and tertiary care levels, so that they have the responsibility of helping population residing in a well-defined geographical area and have suitable transfer linkage with each other. One of the major factors responsible for poor performance in hospitals in the absence of health personnel of all categories posted over there. In this context it became necessary that there must be appropriate of powers to 'Panchayati Raj Institution' (PRI) so that there is local responsibility of the public health care providers and problems relating to poor performance can be sorted out locally.

There will have to be sustained commitment to deliver vital primary health care, emergency life services, facility under the 'National Disease Control Programmes' and the 'National Family Welfare Programme' free of cost to individual based on their to pay.

1.5 Review of Literatures

A number of researchers were made to analyse the supply and cause of disease and spatial pattern of health care studies. However, a few studies previously in the field of medical geography have given momentum towards exploring the issues related with delivery of health care.

India has been a late starter in the field of medical geography. So far only a few studies have been concluded in the field of medical geography. Most of the earlier studies were confined to incidence and spatial distribution of diseases. A significant study of Srinivasan (1984) has been undertaken on micro-scale at village level in Tamil Nadu. This study deals with the awareness and adoption of health care measures. A few researchers Kumra and Singh (1994); Mishra (1988) etc. tried to analyse the utilization pattern of health care services concentrated by primary health centres in rural areas of Uttar Pradesh.

Chakraborti (1972) observed that 54.4 % people of Chiraigaon block of Varanasi district as aware of the medical care programme of PHC's of which 62.2 % availed the services of PHC. Further the consciousness as regard to MCH services was 28 % of which 56.2 % utilized the services of PHC. The area off effective coverage by MCH service varied from 3 to 5 radiuses from PHC. Marwah & Rao (1978) observed that one third health care is provided by through home remedies and one third through folk practitioners in rural area.

Anjaneyulu (1985) in his study of Hyderabad found that individuals in the lower classes were likely to develop the preventive health services. Mishra (1988) in his study of Varanasi district pointed out that in 92.3 % of families deliveries took place at home, whereas only 28.46 % families make visit at PHC's/CHC's for vaccination. Koenig et al. (1988) deliberated the maternal mortality in Matlab, Bangladesh during 1976–1985 and arrived at the conclusion that the introduction of a family planning programme in half of the Matlab study area lead to a moderate but substantial reduction in mortality rate.

Rohde (1990) taking about strategies of PHC perceived that inspite of major investment on PHC/MCH centres only 20 % people used available facilities. He further stated that in India use of antenatal care delivery services etc. were uniformly low. Kumra et. al. (1994) showed a survey of blocks of Varanasi district and detected that out of total 6255 eligible women, 1026 mothers revealed that they have children below 1 year i.e. the children who required child immunization and child check up in first year of their lives of those 1026 mothers, only 35 % knew about free availability of child immunization services through PHC's/CHC's and more pitiable than that only 20.56 % got their children immunized.

Albert et al. (1995) in his research paper entitled "Infectious Disease and GIS," studied applications of geographic information systems that examine spatial aspects of dracunculiasis (Guinea worm disease), LaCrosse encephalitis, Lyme disease, and malaria. For each infectious disease the text followed a sequence that contains an explanation of disease and its transmission chain, the geographic distribution and new statistics, and a review of selected research using geographic information systems. A cross-comparison of conclusions recommends that a targeted approach is more active than broad-based approaches in eliminating or reducing vectors and corresponding rates of infection. This study exhibited the benefit of incorporating elements of human and physical geography into GIS databases used to combat vectored diseases.

Borren et al. (1996) used geographical information systems (GIS) capabilities to find a nested hierarchy of localities for the management of primary health care in West Sussex, England and developed GIS coverage's, which contained key criteria for important local areas, including nodes or focal points of service provision, edges which act as physical or psychological barriers to movement, districts such as official administrative areas and interaction standards such as journey to work, school and family doctor (GP) surgeries.

Richards et al. (1999) studied about the role of GIS for community health planning and provided some common background about the public health marketplace for GIS products, models for establishing GIS within public health, and research challenges connected to GIS software development. They advised that software developers explore the feasibility of forming private-public partnerships with innovative local health departments that have already started to implement GIS. In addition, they also recommend focusing efforts on one (or a few) sentinel local public health issue(s), and developing modules that can be used separately, but that also can be nested together in a diversity of different combinations, depending on a community's specific needs and priorities. In their studied, they emphasized local public health practice with the help of GIS product development is Web-enabled GIS with community-wide access, integrated with community planning tools such as Assessment and Planning Excellence through Community Partners for Health and the Guide to Community Preventive Services.

Albert et al. (2000) discussed role of GIS in Health Service Research. In this paper GIS applications have been organized using the following major divisions: physician distributions; hospitals and other health care facilities; and monitoring, surveillance, and emergency planning. Finally, a discussion followed that critiques the HSR/GIS applications in terms of an intersection of GIS and health services research (HSR). They also presented multiple definitions of GIS and health services research, outlines some general concerns about Geographic Information Systems, and makes a general appraisal of the contribution of this technology to the health of human populations.

Higgs et al. (2009) highlighted the gap between academic health-based applications of Geographical Information Systems (GIS) and their everyday use within the

UK National Health Service (NHS). They provided examples of the operational benefits ensuing from using GIS in a range of health care applications whereas acknowledging the limited use of such technologies in strategic health tasks. They decided by re-emphasizing the importance of using GIS in strategic health planning contexts in the light of both recent health-care restructuring and new technological developments in the health service and also studied the impact of distance, in conjunction with non-spatial factors in retrieving and utilizing healthcare.

Rezaeian et al. (2007) focused on the issue of disease mapping, i.e., the production of clean maps of the geographical distribution of disease (incidence or prevalence) and also advised the representation of disease-incidence data can vary from simple point object maps for cases and pictorial representation of counts within tracts, to the mapping of estimates from complex models asserting to designate the structure of the disease events and described the range of mapping methods from simple representations to model-based forms.

Murad (2004) described the approaches to geographical analysis and focuses on cartographic operations, exploratory spatial data analysis, and ecological studies. The use of GIS and cartographic operations is strongly endorsed by them to bring an added dimension to public health intelligence. Their importance is on data on healthcare factors that may be available through routine health information systems. Such data may comprise numbers of beds available, numbers of consultants and other specialist staff, bed occupancy rates or delivery of specialist services etc.

Hoogh et al. (2001) developed and defined buffer zones around source- a point, line, or area-to recognize allegedly exposed and unexposed groups in Teesside area. A single buffer might be used to divide populations into exposed and unexposed groups or multiple buffers can be created to define degrees of exposure. Standard epidemiological techniques are used to associate disease rates in the various groups with control, if suitable, for confounding by other factors.

Rohan (2002) gathered GIS data from the agencies and developed within a GIS to build up a set of spatial databases of available services, location of users and additional geo-demographic and topographic information. The output from this system development was presented in turn at workshops with agencies associated with short-term care planning as well as users to help assess their perspectives on the potential use and value of GIS. A renewed emphasis on a planned approach to health care coupled with integrated/ joint working with social care creates a need for new approaches to planning.

Sankar and Satish (2003) developed Spininfo Health Map for Karnataka Health Systems. They implemented custom GIS application which would be an interactive spatial analysis tool enabling the Health Officer to perform re-districting, re-locating health jurisdictions for effective utilization of health infrastructure. They Rationalized Health Services Infrastructure to serve allocated population and did spatial analysis on disease surveillance and monitoring and other health indicators, and focus on a particular area.

McLafferty (2003) discussed in her literature on GIS and health care. She considered the use of GIS in analyzing health care need, access, and utilization; in planning and evaluating service locations; and in spatial decision support for health care delivery. The adoption of GIS by health care researchers and policymakers will depend on access to integrated spatial data on health services utilization and outcomes and data that cut across human service systems.

Barik (2004) prepared the integrated health management system (IRHMS) of Ranchi district. This system may be helpful in understanding the function of GIS in rural health management. The micro-level studies of health care facilities are positively related with the preventive and curative health care. He also perceived availability of health care facilities, literacy and educational levels, occupation in non-primary sector and income has also positive associations with the indicators of health status.

Saxena et al. 2012 also worked on malaria situation for priority control at micro level in Ranch using GIS and statistical methods. In this study A geographical information system (GIS) based retrospective study using spatial statistical tools was initiated in 328 subcentres of 14 primary health centres (PHCs) of the district using malaria epidemiological data of 3 years (2007–2009) to identify spatial distribution pattern of *Plasmodium vivax (Pv)* and *Plasmodium falciparum (Pf)* occurrence, delineation of hot spots and to map directional distribution trend of *Pf* spread to help formulate evidence-based policy and to prioritize control during 2011.

Chung et al. (2004) observed the extent to which health studies, mostly in public health and epidemiology, used geographical information systems (GIS) and identified a wide range of tools they used—ranging from geocoding through simple buffer or overlay functions to spatial query functions. However, their study tend to depend on tools outside of GIS for spatial statistical analyses and may reflect a lack of spatial statistical tools that are appropriate for health researchers whose data are rather geographically combined count data than continuous data.

Luis (2004) assembled a geographic information system (GIS) to relate the 2000 census population (demand) with an inventory of health facilities (supply). It considers the fairness in access to health care by Costa Rica and the influence on it by the ongoing reform of the health sector and uses traditional measurements of access based on the distance to the closest facility and suggests a more comprehensive index of convenience that results from the aggregation of all facilities weighted by their size, proximity, and physiognomies of both the population and the facility.

Murad (2004) studied local health care planning in Saudi Arabia. They discuss how Geographical Information Systems can be used to care health planners on a micro-scale. They assessed the issue that affects local health care planning which include monitoring of catchments area and facilities management and also describes GIS and its possible uses in the health care field. In this application, three sets of GIS models have been produced. These are catchment's area, patient profile and patient distribution and patient flows models. The created GIS models are produced to help local health planners in their health care decision output.

Mandy et al. (2004) analysed the spatial and temporal patterns of oviposition behaviors, a web-based Geographical Information System (GIS) for dengue fever surveillance equipped with query and buffer analysis functions. In this study spatial statistical analysis can be applied to help, identify and visualize potential risk areas or hot spots affected by Aedes albopictus.

Chapelet and Lefebvre (2005) introduced setting-up of a GIS technique of the Delhi agglomeration for the purpose of selecting representative samples of urban areas and health infrastructure for fieldwork investigations and to better assess the socio-spatially dynamics taking place in the Indian capital. Using GIS technique in their work, they selected samples of health infrastructure, in particular those distributing medicines, for fieldwork survey according to their location in the urban agglomeration and also selected urban areas according to their (socio-economic, demographic, as well as according to the healthcare delivery system) for further investigation (survey of households).

Zhan et al. (2006) presented a simple spatial search tool-GIS EpiLink- that can be used to link environmental and health data when distance between an environmental site and the location of the maternal address of a case or control is used as a proxy for exposure. The tool was used in a research scheme and positively formed the necessary data for epidemiological analyses and should be very valuable to epidemiologists whose research requires the link of environmental and health data based on distances between environmental sites and the locations of maternal addresses of cases or controls. Such a tool could also be used to address community concerns about apparent excesses of birth blemishes and other adverse reproductive outcomes around hazardous waste sites and industrial facilities.

Peng et al. (2006) proposed a conceptual model simulating the spatio-temporal dynamics of infectious diseases and based the model on the knowledge of the interrelationship among the source, media, and the hosts of the disease. With the endemics data of schistosomiasis in Xichang, China, they confirmed that the conceptual model is viable and familiarized how remote sensing and geographic information systems techniques can be used in support of spatio-temporal modeling and also related the different effects caused to the whole population when choosing different groups of people for schistosomiasis control. Their exemplified the importance of such a modeling tool in supporting spatial decisions.

Prashanthi Devi et al. (2007) used remote sensing and GIS techniques to highlighted Socio-Demographic Determinants of malaria in highly infected rural areas of Tamil Nadu. In this work, a group of highly endemic villages and an uninfected village with related topography were surveyed for their geographical features, distance from house to breeding sites, the socio economic living condition of a community (including the construction type of houses), and the use of anti-malaria measures.

Ronit Peled et al. (2006) developed an effectual tool for quality declaration and chronic disease management using a Geographic Information System (GIS) and produced a list of six markers for insufficient pharmaceutical treatment of childhood asthma from the Israeli clinical rules. They used this list to observe the drug dispensing registry to differentiate asthmatic children who received unsatisfactory treatment and to evaluate their health care utilization and bad outcomes: emergency room visits and hospitalizations and they also formed thematic maps on which located clinics with a high percentage of children for whom the treatment provided was not in adherence with the clinical rules.

Black et al. (2007) designated two possible statistical methods integrated with GIS techniques for computing physical accessibility to health care. For both of the methods the objective is to deliver tools that could be used in countries, and more predominantly in developing countries, where the need to access health care is one of the key elements for reducing the burden of diseases. The results of these methods are used for cost effectiveness analysis, population coverage estimates as well as for resource planning within countries. They also deliberated the benefits for better health planning and policy development through the use of GIS before describing possible enhancements to the models in the future.

Bhunia et al. (2012 and 2013) lead a study on kala-azar, Vaishali district, Bihar (India) and GIS techniques are used to analyse the distribution pattern and geographic location of the disease as well as its connection between vector, hosts and reservoir. The distribution of kala-azar cases was obtainable on kala-azar density maps which were calculated for each PHC based region. Data analysis has also revealed that above 5 % of the total kala-azar cases reported in Bihar from the district from 1991 to 2005. In their work, they highlighted the role of GIS in monitoring and management of kala-azar control programme.

Joyce (2009) explored four thematic areas (the ontological, power, functionality and collaboration discourses) to comprehend how GIS are perceived and appreciated by public health decision-makers through analysis of data captured in semi-structured interviews. The findings advise that although GIS are observed as useful tools to inform decision-making, they are in no way a solution for practice. He also highlighted about 'participants' concerns that GIS outputs can possibly be misinterpreted or used speciously might partly explain resistance to their use and therefore, likely to be most operative in decision-making when applied in a multi-disciplinary framework to simplify sharing of data, knowledge and capability across the public health landscape.

1.6 Objective of the Study

The main aims of the study are:

(a) To analyse the population characteristics.
(b) To attain distributional pattern of health care facilities using GIS.
(c) To evaluate the status of vector borne diseases and malaria mapping using statistical methods with GIS in the study area.
(d) To estimate the utilization pattern of Health care facilities.
(e) To measure the percolation of awareness of the people towards health services available at PHC's.

1.7 Hypotheses of the Work

The hypotheses of the study are selected in the following lines:

(a) The distance is a vital factor in the utilization of health care facilities i.e. as the distance from the government health centres increases the level of utilization decreases.
(b) As the distance from the health facilities increases the malaria area percentage also increases.
(c) Information value (Infoval) method is designated as an optimum model for malaria susceptibility zonation.

1.8 Data Used and Methodology for the Study

In this study, Survey of India (SOI) topographical is used to extract the study area and development block boundaries of Varanasi district. Extraction of study area boundary in shape file and other vital inputs used in this study eg. roads, location of health care facilities etc. are digitized by ARC GIS-9.3 software. Village map is also used in this study in which correct existing government health care facilities i.e. location of PHC's, CHC's etc. are located using Global Positioning System (GPS), which provide accuracy in 3–5 mm. Remote sensing data i.e. IRS-1C LISS III of 2008 is also used for this study for preparing Land Use/Land Cover information of the area using image processing supervised classification technique. Normalized Difference Vegetation Index (NDVI) is also calculated to highlight the vegetation and water bodies. All these information are interpreted on ERDAS Imagine-9.2 image processing software. This information is used as important parameters to developed malaria model and was very valuable to relate their impact on malaria disease. Climate data i.e. rainfall, temperature, census data and primary data were also used for the malaria disease analysis. Primary data was collected from 800 respondents of selected 16 villages (2 villages from each development blocks) in the rural area of Varanasi district to know about the utilization of health care facilities. Varanasi city is not considered for this purpose because here many private and government hospitals are available where people find quite good health facilities in comparison to people living in the rural area. To estimate the status of vector born disease, Varanasi city area is also comprised with eight development block.

All the indicators and GIS inputs were used for statistical analysis to develop a malaria model of the Varanasi district using multiple linear regression, heuristic and information value (infoval) methods. Optimum model of malaria susceptibility index (MSI) and malaria susceptibility zones (MSZ) is used to calculate hospital requirement index (HRI) for health care planning. GIs based network analysis was also used to show the catchment area of PHC's, and CHC's and to calculate the shortest path between one village to a PHC's or from one PHC's to another PHC's/ CHC's.

1.9 Organization of Chapters in This Study

This study is organized into eight chapters. This chapter is devoted to conceptual framework of the study, objectives and methodology of the study. The review of literature is also included in this chapter. Chapter 2 deals the study area profile. Chapter 3 describes role of integration of GIS with population data and to assess the population characteristics in the area. The spatial distribution and analysis of health care facilities using GIS is discussed in Chap. 4. This chapter provides bright vision of the spatial analysis of availability of health facilities at different point of time, the block levels as well as in terms of area and population. Buffer area analysis and network analysis are also used in this chapter to the safe and unsafe villages and also to calculate the shortest path between a village to primary health centre's (PHC's) and community health centre's (CHC's). Chapter 5 provides the information about distribution of vector born diseases with special reference to Malaria diseases in the study area. Chapter 6 deliberate the role of GIS in malaria susceptibility mapping using various statistical methods. Many important indicators are used to develop a model for malaria affected place. This section also gives comprehensive description of disease distribution to the calculation of rates, interpolation methods malaria mapping and modeling from a statistical perspective using GIS techniques. Comparison of statistical methods for optimum model of malaria susceptibility index (MSI) and malaria susceptibility zone (MSZ) is also calculated in this chapter. Chapter 7 examines the utilization pattern of existing health care facilities in eight blocks of Varanasi district. For this purpose, 800 people from 16 villages were interviewed through questionnaire. Each individual is questioned general biographical information, opened ended view questions related to their utilization of health care facilities. These answers are then coded and statistically analysed and presented in SPSS software. Chapter 8 illustrates the problems and possible GIS and statistical methods of planning parts for health care in the study area.

References

Acheson, D. (1988). *Public health in England: The report of the committee of inquiry into the future development of the public health function* (Cmnd 289). London: HMSO.

Albert, D. P., Gesler, W. M., & Wittie, P. S. (1995). Geographic information systems and health: An educational resource. *Journal of Geography, 94*(2), 350–356.

Albert, D. P., Gesler, W. M., & Levergood, B. (2000). *Spatial analysis, GIS, and remote sensing applications in the health sciences*. Chelsea: Ann Arbor Press.

Ali, M., Rasool, S., Park, J.-K., Saeed, S., Ochiai, R., & Nizami, Q. (2004). Use of satellite imagery in constructing a household GIS database for health studies in Karachi, Pakistan. *International Journal of Health Geographics, 3*(1), 20.

Anjaneyulu, G. (1985). A study of polio myelities case admitted in government hospital, Hyderabad. *Indian Journal of Public Health, 32*(2), 97.

Annual Report. (1980). *Ministry of Health and Family Welfare (MOHFW)*. New Delhi.

Annual Report. (1987). *National cancer registry programme*. New Delhi: ICMR.

Barik, V. K. (2004). GIS in health management. *Proceeding of Recent Trend in Geoinformatics*, BIT Mesra, 45–50.

Beck, L. R., Lobitz, B. M., & Wood, L. B. (2000). Remote sensing and human health: New sensors and new opportunities. *Emerging Infectious Diseases, 6*(3), 217–227.

Bhunia, G. S., Chatterjee, N., Kumar, V., Siddiqui, N. A., Mandal, R., Das, P., & Kesari, S. (2012). Delimitation of kala-azar risk areas in the district of Vaishali in Bihar (India) using a geo-environmental approach. *Mem Inst Oswaldo Cruz Rio de Janeiro, 107*(5), 609–620.

Bhunia, G. S., Kesari, S., Chatterjee, N., Kumar, V., & Das, P. (2013). Spatial and temporal variation and hotspot detection of kala-azar disease in Vaishali district (Bihar), India. *BMC Infectious Diseases, BMC series, 13*, 64. doi:10.1186/1471-2334-13-64.

Black, M., Ebener, S, Aguilar, P. N., Vidaurre, M., & Morjani, Z. El. (2007). Using GIS to measure physical accessibility to health care, WHO, Pan American Health Organization.

Borren, N., Moon, G., & Jones, K. (1996). Defining localities for health care planning: A GIS approach. *Social Science Medicine, 42*(6), 801–816.

Cano, J., Descalzo, M., Ndong-Mabale, N., Ndongo-Asumu, P., Bobuakasi, L., & Buatiché, J. (2007). Spatial and temporal variability of the Glossina palpalis palpalis population in the Mbini focus (Equatorial Guinea). *International Journal of Health Geographics, 6*(1), 36.

Chakraborti, M. K. (1972). *A study of working health centre, Chiraigaon, Varanasi*. M.D. Thesis (P.S.M.), Banaras Hindu University, pp. 1–226.

Chapelet, P., & Lefebvre, B. (2005). *Contextualizing the urban health care system: Methodology of developing a geodatabase for Delhi's health care system*. CSH Occasional Paper N°, 1–82.

Chung, K., Yang, D. H., & Bell, R. (2004). Health and GIS: Toward spatial statistical analyses. *Journal of Medical Systems, 28*(4), 349–360.

Cromley, E., & McLafferty, S. (2002). *GIS and public health*. New York: The Guilford Press.

Curtis, S. E., & Taket, A. R. (1989). The development of geographical information systems for locality planning in health care. *Area, 21*(4), 391–399.

Daly, P. (1993). Navstar GPS and GLONASS: Global satellite navigation systems. *Electronics and Communication Engineering Journal, 5*(6), 349–357.

Datta, S. P., & Kale, R. V. (1969). An operation research study in primary medical care in Pondichery. *Indian Journal of Preventive Social Medicine, 1*, 65–72.

de Hoogh, K., Briggs, D., Samantha, C., & Bottle, A. (2001). Using GIS for environmental exposure assessment: Experiences from the small area health statistics unit. In R. Maheswaran & M. Cragilla (Eds.), *GIS in public health practice* (pp. 109–124). New York: CRC Press.

DevenhaLL, b. (2002). Health geography. *ESRI magazine*, June-Sept. 2002.

Devi, P., Ranganathan, C. R., & Balasubramanian, S. (2007). *Socio-demographic determinants of malaria in highly infected rural areas: Regional influential assessment using GIS* (GIS for health and the environment, pp. 195–205).

Donald, P. A., Wilbert, M. G., & Barbara, L. (2005). *Spatial analysis, GIS and remote sensing application in the health*. Chelsea: Ann Arbor Press.

Dwolatzky, B., Trengove, E., Struthers, H., McIntyre, J., & Martinson, N. (2006). Linking the global positioning system (GPS) to a personal digital assistant (PDA) to support tuberculosis control in South Africa: A pilot study. *International Journal of Health Geographics, 5*(1), 34.

Escobar, F., Williamson, I., Waters, E., Green, J., Hugo, G., & Rudd, C. (1997). *The role of GIS in the management of primary health care services*. AURISA 97 – The 25th annual conference of the Australasian Urban and Regional Information Systems Association, Christchurch, New Zealand.

Flabault, D. (1972). The training of rural health personnel. W.H.O. *The Chronicle, 26*, 243–249.

Ghosh, B. N., & Mukharjee, A. B. (1989). An analysis of health services coverage of primary health centre in West Bengal. *Indian Journal of Public Health, 33*(1), 12–19.

Glass, G. E. (2000). Update: Spatial aspects of epidemiology: The interface with medical geography. *Epidemiologic Reviews, 22*(1), 136–139.

Goel, S. L. (1984). *Report of planning commission, public health, administration*. New Delhi: Sterling Publishers.

Goel, S. L., & Rajneesh, S. (2007). *Management and techniques; principle and practice* (Rev. ed.). Management of the employees' health. (pp. 753–764). New Delhi: Deep and Deep Publication.

Higgs, G. (2009). The role of GIS for health utilization studies: Literature review. *Health Service Outcomes Research Methodology, 9*, 84–99.

Hunter, J. M. (1974). The challenge of medical geography. In J. M. Hunter (Ed.), *The geography of health and disease* (Studies in geography series no. 6, pp. 1–31). Chapel Hill: Department of Geography, University of North Carolina.

Joyce, K. (2009). To me it's just another tool to help understand the evidence: Public health decision-makers' perceptions of the value of geographical information systems (GIS). *Health & Place, 15*, 831–840.

Khan, M. E. (1980). *People's perception about family planning in India: A study of Andhra Pradesh and Bihar*. New Delhi: Concept Publishing Co.

Koenig, A. M., Fauveau, V., Chowdhury, A. I., Chakraborty, J., & Khan, M. A. (1988). Maternal mortality in Matlab, Bangladesh: 1976–1985. *Studies in Family Planning, 19*, 69–80.

Kumra, V. K., & Singh, J. (1994). Assessment of Community Attitude regarding the services of primary health centre: A medical geography study. *Indian Journal of Preventive & Social Medicine, 25*(3/4), 118–126.

Luis, R. B. (2004). Spatial access to health care in Costa Rica and its equity: A GIS-based study. *Social Science & Medicine, 58*, 1271–1284.

Mahmoud, M. S. A., Abdullah, S. M. A., & El-Tayeb, N. M. (2014). Improve of health care systems in subareas of SaudiArabia by GIS. *Control Theory and Informatics, 4*(5), 6–57.

Mandy, A., Milton, C., & Mandy, P. (2004). Professional stereotyping and interprofessional education. *Learning in Health and Social Care, 3*(3), 15470.

Marwah, S. M., & Rao, N. S. N. (1978). *Health profiles of Chiraigaon block*. Varanasi: Department of P.C.M., I.M.S., B.H.U.

McLafferty, S. L. (2003). GIS and health care. *Annual Review on Public Health, 24*, 25–42.

Meade, M. S., & Earickson, R. J. (2000). *Medical geography*. New York: Guilford Press.

Meade, M., Florin, J., & Gesler, W. (1988). *Medical geography*. New York: The Guildford Press, 340pp.

Mishra, C. P. (1988). A study of Community practices regarding MCH care in rural areas of Varanasi. *India Journal of Public Health, 32*(2), 92.

Mohler, H. (1978). *Primary health care justice in health* (p. 3). Geneva: World Health.

Murad, A. A. (2004). Creating a GIS application for local health care planning in Saudi Arabia. *International Journal of Environmental Health Research, 14*(3), 185–199.

Nicol, J. (1991). Geographic information systems within the national health service: The scope of implementation. *Planning Outlook, 34*(1), 37–42.

Nipada, R. (2005). GIS and epidemiology. *Journal of Medical Association, Thai, 88*(11), 1735–1738.

Noack, H. (1980). *Medical education and primary health care* (pp. 132–133). Baltimore: University Park Press.

Pacione, M. (2014). *Medical geography: Progress and prospect* (Routlegde revivals). London: Routlegde.

Park, K. (1995). *Park's textbook of preventive and social medicine* (p. 14). Jabalpur: Mis Banarsidas Bhanot.

Peled, R., Reuveni, H., Pilskin, J. S., Beneson, I., Hatna, E., & Tal, A. (2006). Defining localities of inadequate treatment for childhood asthma: A GIS approach. *International Journal of Health Geographics, 5*, 3. doi:10.1186/1476-072X-5-3.

Pellerin, C. (2006). *United States updates global positioning system technology*. Retrieved July 11, 2008, from http://www.america.gov/st/washfileenglish/2006/February/20060203125928lcnirellep0.5061609.htm

Peng, G., Xu, B., & Song, L. (2006). Remote sensing and geographic information systems in the spatial temporal dynamics modeling of infectious, diseases. *Science in China Series C: Life Sciences, 49*(6), 573–582.

Rezaeian, M., Dunn, G., Leger, S. S., & Appleby, L. (2007). Geographical epidemiology, spatial analysis and geographical information systems: A multidisciplinary glossary. *Journal of Epidemiology and Community Health, 61*(2), 98–102.

Richards, T. B., Croner, C. M., Rushton, G., Brown, C. K., & Fowler, L. (1999). Geographic information systems and public health: Mapping the future. *Public Health Reports, 114*, 359–373.

Rohan, F. (2002). Assessing the applicability of GIS in a health and social care setting: Planning services for informal carers in East Sussex. *England, Social Science & Medicine, 55*, 79–96.

Rohde, J. E. (1990). A strategy for strengthening PHC. *Indian Journal of Community Medicine, 15*(4), 185–194.

Rosicky, B. (1965). *Medical geography, lecture XIV, Courses in advanced entomology N.I.C.D.* New Delhi

Sankar, K. N., & Satish, S. (2003). Spinfo health map- a health GIS application. www.gisdevelopment.net/application/health/overview/mi0318.htm

Saxena, R., Nagpal, B. N., Das, M. K., Srivastava, A., Gupta, S. K., Kumar, A., Jevaseelan, A. T., & Baraik, V. K. (2012). A spatial statistical approach to analyze malaria situation at micro level for priority control in Ranchi district, Jharkhand. *Indian Journal of Medical Research, 136*(5), 776–782.

Senthil, J., Rajeswari, S., Vadivel, S., & Malarvizhi, F. (2004). Scenario of dengue and health care utilization in Nagapattinam town, Tamilnadu, India. *International Journal of Development Research, 4*(11), 2547–2552.

Srinivasan, S. (1984). *A primary health centre in Tamil Nadu*. Kurukshetra, p. 15.

Srivastava, A., & Nagpal, B. N. (2000). Mapping malaria. *GIS Development, IV*(6), 28–31.

UNICEF (1978). Annual Report, New York.

Verhasselt, Y. (1993). Geography of health: Some trends and perspectives. *Social Science and Medicine, 36*(2), 119–123.

World Health Organization (WHO). (1948). Constitution of the World Health Organization (Vol. 2, p. 100). Geneva: WHO.

World Health Organization (WHO). (1978). *International conference on primary health care, Alma-Ata, USSR.*

Zhan, F. B., Brender, J. D., Han, Y., Suarez, L., & Langlois, P. H. F. (2006). GIS-EpiLink: A spatial search tool for linking environmental and health data. *Journal of Medical Systems, 30*, 405–412.

Chapter 2
Applying Remote Sensing and GIS in Study of Physical and Cultural Aspects of Varanasi District

Abstract The city of Varanasi (Benaras) lies on a bend of Ganga river (Ganges), one of the largest rivers of the World. In this chapter, various physical i.e. geological, physiographic, drainage, climatic characteristics, condition of ground water, land use etc, as well as cultural aspect like transportation and communication facilities, solid waste generation etc. through remote sensing and GIS are very well presented. On the basis of relief variation, geology and drainage characteristics the area has been divided into three main physiographic divisions i.e. Upper Ganga-Varuna Plain,Varuna-Gomati Interfluence and Ganga-Varuna Interfluence. In this study it is found that the current water supply system is not adequate to supply to around 1.6 million people in the city. There is a substantial gap between the recommended norm and the actual water supply. Land use land cover (LULC) map is prepared using remote sensing data (IRS-1C/1D-LISS-III) of 2008, in which five important classes are delineated i.e. agricultural field, fallow land, vegetation, built-up area, and water bodies. In Varanasi district there are 85 Ganga ghats with typical structure on the river Ganga which are frequented by a large number of pilgrim's everyday for taking a holy dip in river Ganga. The city along the Ghats has facing the severe serious problems of solid waste management due to congested lanes. These physical and cultural aspects are directly and indirectly used in the main themes of the study.

Keywords Varanasi • Ganga river • GIS • Remote sensing data • IRS-1C LISS-III • Land use land cover • Solid waste management

2.1 Locational Extent of the Area

The city of Varanasi (Benaras, Benares) lies on a bend of river Ganga (Ganges), one of the largest rivers of the World. It is the oldest living city in the world. Varanasi district, covering between 25°10′ and 25°37′ N latitude and 82°39′–83°10′ E, lies in eastern Uttar Pradesh of India (Rai and Mohan 2014). The city cover over an area of 1454.11 sq. km, it has a total population of 1,878,100 (as year 2001 census report) persons (976,045 males and 902,045 females). It is bounded by Bhadohi district in the west, Jaunpur district in north-west, Ghazipur district in the north and north east,

P.K. Rai, M.S. Nathawat, *Geoinformatics in Health Facility Analysis*,
DOI 10.1007/978-3-319-44624-0_2

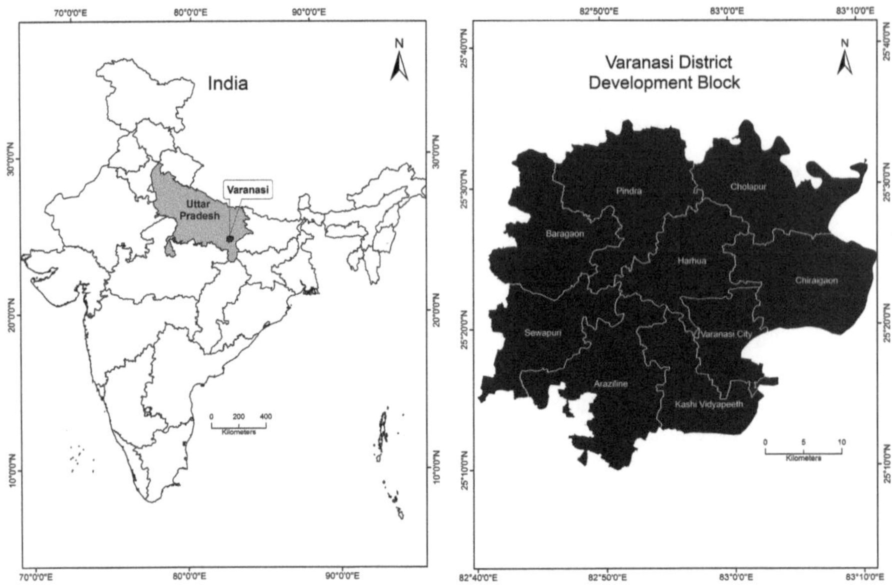

Fig. 2.1 Location map of the study area

Chandauli district in the east and Mirzapur district in the south (Fig. 2.1). The Ganga river forms its natural boundary in the east and south east while northern boundary is noticeable by the river Gomti. The Grand Trunk Road (NH2) passes through the city linking the study area with prominent northern and eastern cities of the country. From administrative point of view, the district comprises two tahsils namely, Pindra and Varanasi Sadar which are further sub-divided into eight development blocks namely Baragaon, Pindra, Cholapur, Chiraigaon, Harhua, Sevapuri, Araziline and Kashi Vidyapeeth comprising of 1336 villages. This holy city is also known as Kashi (the city of light). The ancient city of Varanasi is inimitable in its architectural, artistic and religious expressions of traditional Indian culture which is existing even today. The city has been well-known for knowledge along with art works, cottage and textiles industries ever since pre-Buddhist times. Silk weaving and sari making, wood and terracotta handicrafts, toy making, particular painting forms, etc., are vivid the example of continuing historic-cultural tradition.

The numerous Hindu temples and the ghats of river Ganges are the places of interest in Varanasi. The Vishwanath temple is considered as the eternal abode of Lord Shiva on earth. Another place to be visited is Sarnath, an ancient Buddhist site, Banaras Hindu University (A Central University), a seat of oriental and modern learning can be counted as a place to be visited. Study area as viewed on IRS-1C LISS III remote sensing data is shown in the Fig. 2.2.

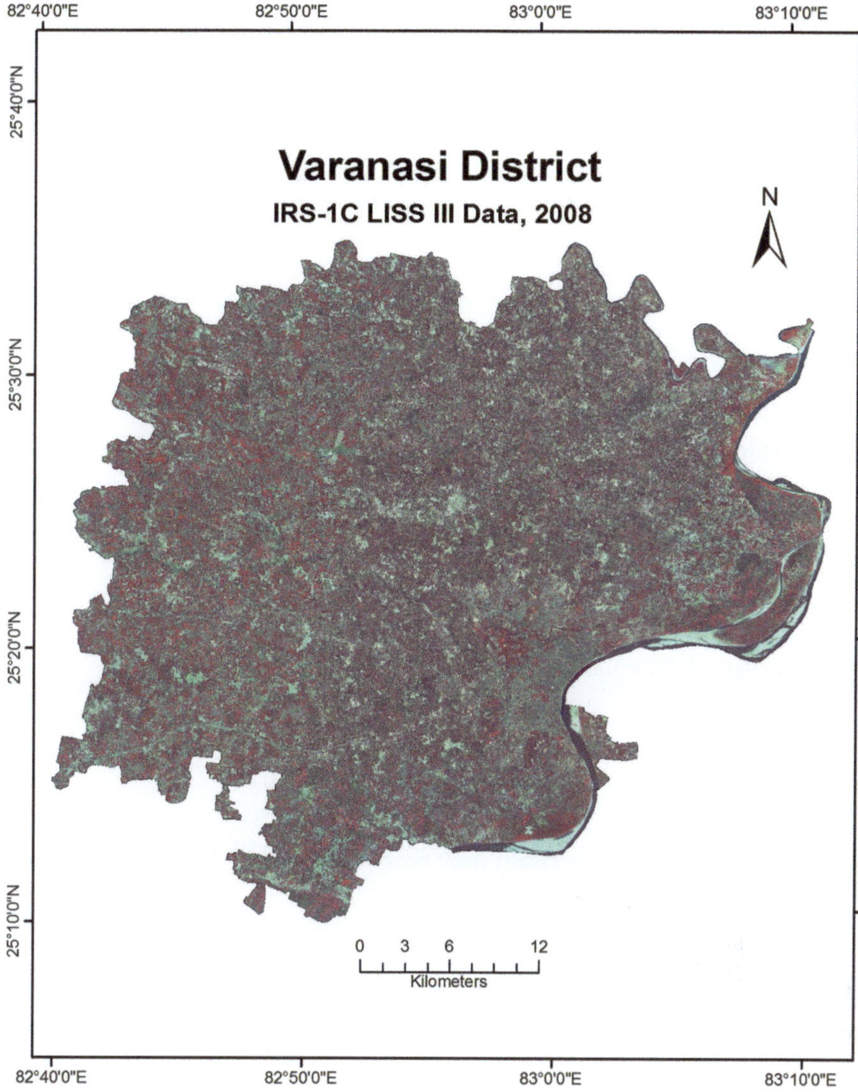

Fig. 2.2 Varanasi district as viewed on IRS-1C LISS III data, 2008

2.2 Geology of the Area

Varanasi is part of the Ganga flood plain, underlain by the alluvial deposits of quaternary age. The older alluvium of the Ganga valley from slightly elevated land surface of the area. This deposit is supposed to be middle to upper Pleistocene and lies well above the highest flood level of the rivers of the area. The newer alluvium generally occupies the lower elevation and is constrained to the flood plains of the present day river along their channels and consists of fine to medium grained sands, silts and minor amount of clays. However, there is not always a clear delineation line between older alluvium and newer alluvium. The Older alluvium is dark coloured and generally rich in concretions and nodules of impure calcium carbonate. These concretions are of various shape and size and are known as 'Kankar'. The Varanasi older alluvial plain represents the oldest geomorphic surface occurring at highest tectonic level in the Gangetic plain. The study area placed between the Ganga in the south and the Gomti in the north with the river Varuna flowing in central part imprinting upon the geology of the areas as it's comprises of inter-bedded layers of sand, silt and clay sometime associated with 'Kankar' and clay (Rai and Mohan 2014).

2.3 Physiographic Divisions

On the basis of relief variation, geology and drainage characteristics the area has been divided into three main physiographic divisions:

(i) Upper Ganga-Varuna Plain,
(ii) Varuna-Gomati Interfluence and
(iii) Ganga-Varuna Interfluence.

2.3.1 Upper Ganga-Varuna Plain

Upper Ganga-Varuna plain is extent over nearly 31.05 % of the district. Varanasi district stands on the left blank of river Ganga and lies in the heart of the middle Ganga Valley. The river Varuna reaches from western side along the Dhaukalganj lying in Bharagaon development block and drains in Pindra and Harahua development blocks. The area designates a near level surface having alluvium deposits may be separated into: (a) low land and (ii) upland. Low land known as 'Tarai' is the recent creation of river floods and afresh deposited alluvium soil, while 'Bhangar' (upland) lying outside the flood limits is ended by older alluvium (Rai and Mohan 2014).

2.3.2 Varuna-Gomati Interfluence

The Varuna-Gomati interfluence covers about 40.84 % of the total area of the district. It is yet again classified into two units: (i) upper region (Uparwar) and lower land (Tarai). The upper region recognize in the northern and western parts of the district where no flood arises even in the rainy season also. The low land (Tarai) has further been classified into: (i) flooded areas along the river Varuna, and (ii) flooded areas along the river Gomati and its tributaries. Towards the side of the river Varuna in Varuna-Gomati interfluence, the village namely Sirisi, Barthara, Kalan, Parnapur, Amauli, eastern part of Jalhuapur, Mustafabad, Ambapur, Dhobahi, Shidasa Goberha, Rampur etc. are regularly flooded during the monsoon season locally called as tari villages alluvium (Rai and Mohan 2014).

The eastern part of Bela, Azagara, Bartharakhund and Rampur etc. are flooded by the river Nand a tributary of the river Gomati in the period of heavy downpour. Northern part of Rajwari and Tekuri, western part of Dhauradhara and eastern part of Azgara etc. are also inundated. Hathi nala another branch is frequently flooded in rainy season causing much damage of the life and property to the villages Azgara, Benipur etc. A very small area of the villages Akatha, Baraipur and northern part of Hiramanpur covered by Narokhar, Tal along the river Varuna is also inundated in rainy season (Rai and Mohan 2014).

2.3.3 The Ganga-Varuna Interfluence

The region covers the areas of Sewapuri, Araziline and Kashi Vidyapeeth development blocks lying in the western and southern part of the district. It covers about 28.11 % of the total areas of the district. It can further be sub-divided into two parts:

(i) The Upper Plain of Ganga-Varuna Interfluence, and
(ii) The Flood Zone

Over all, Varanasi district establishes its 4.79 % of the total land along the river banks to be flooded frequently in monsoon season causing damage to human life and property both. The uplands are designated as Bhangar and low lying areas liable to inundation in the rainy period are known as Khadar (known as Tarai) with newer alluvium deposited by the rivers during the floods (Rai and Mohan 2014).

Gemorhological map of the study area is prepared through remote sensing satellite data and the area are divided into 5 important geomorphological class i.e. alluvial plain, point bars, sand bars, river island with alluvial deposits and river island with sand deposits (Fig. 2.3). The study area is associated to the flood plain deposit, so that no other geomorphological features are recognized.

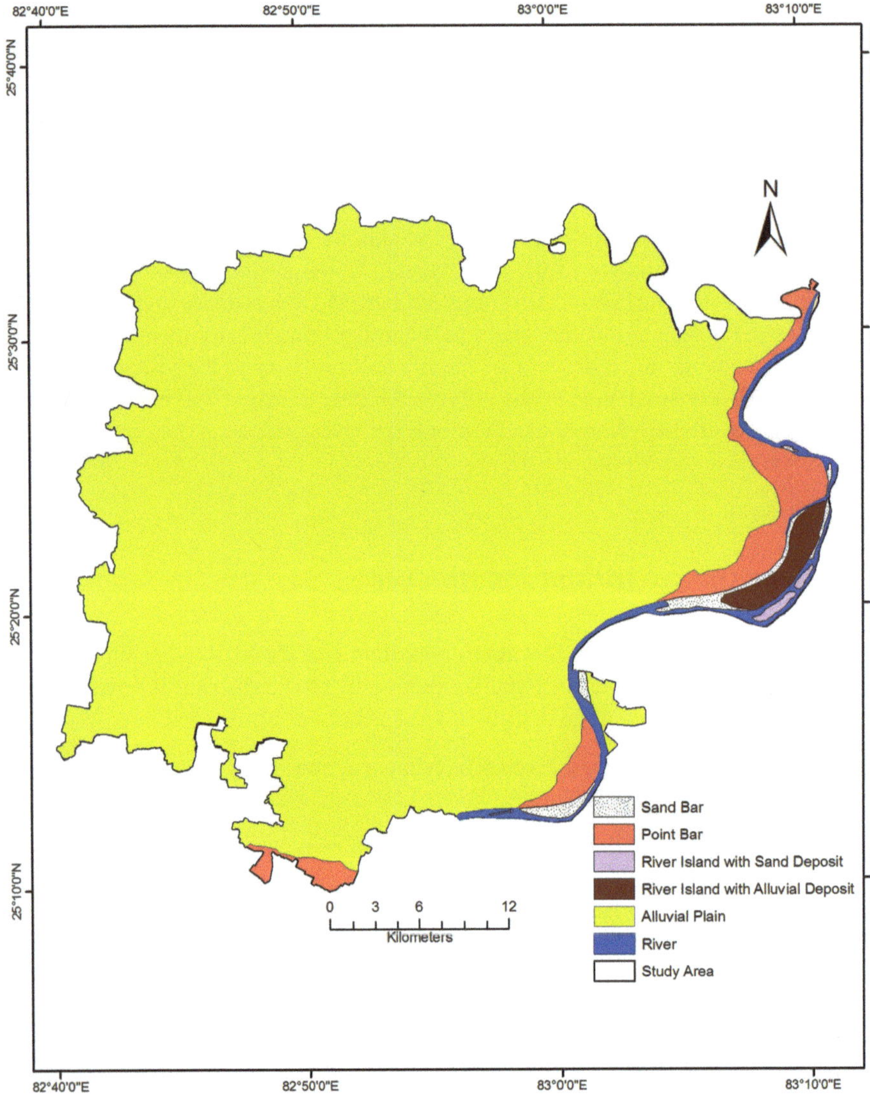

Fig. 2.3 Geomorphological features in Varanasi district

2.4 Drainage and Water Ponds

The drainage plays very important role in shaping the physical characteristics of the area under study. The river Ganga with its tributaries forms the drainage pattern of the study area. The river Ganga is an important source of drinking water as well for irrigation also. The existing water bodies are extracted through IRS-1C LISS III remote sensing data of 2008 (Fig. 2.4).

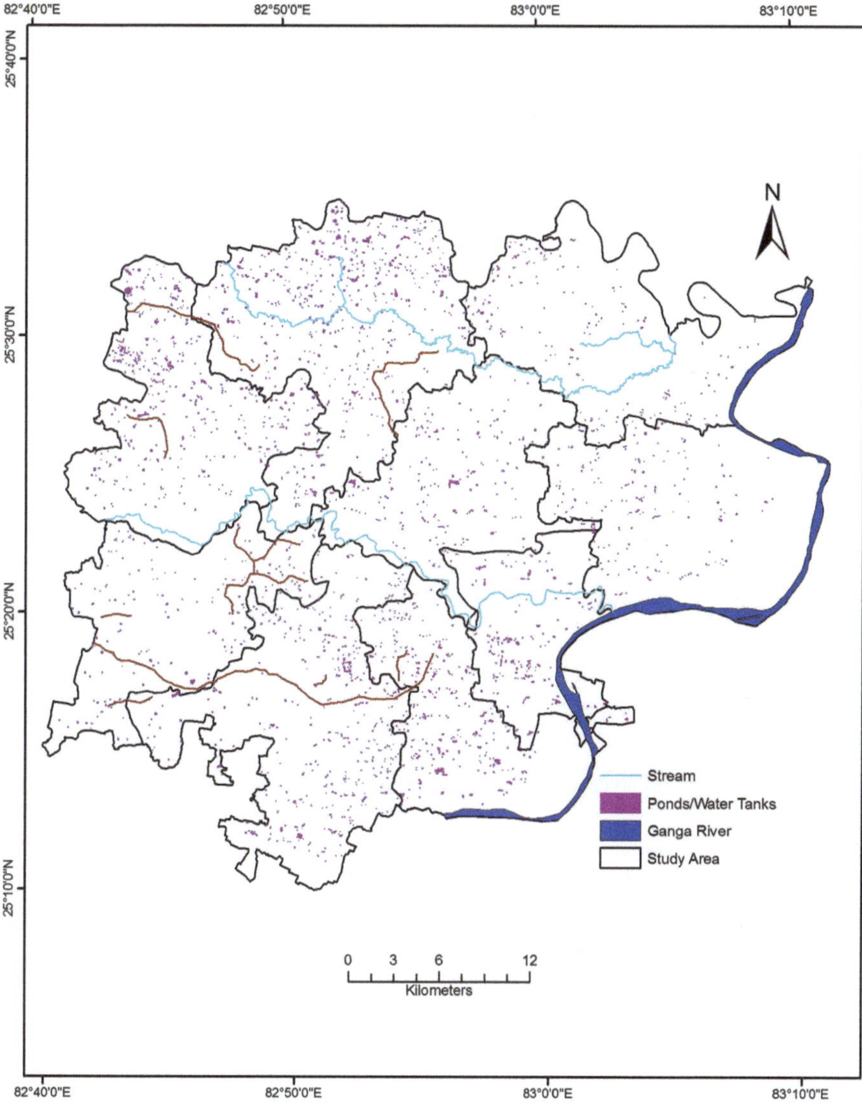

Fig. 2.4 Existing water bodies in Varanasi district (Extracted from IRS-1C LISS-III Data, 2008)

2.4.1 The Ganga River

The River Ganga is the main drainage line of the study area which forms the southern boundary of Kashi Vidyapeeth development block is flowing from the eastern side of the Varanasi city. The Ganga touches the Varanasi district at village Tandi where a stone quarry is situated on the confluence of Subba Nala and the Ganga.

The river flow towards the east in Araziline development block and takes a turn towards the north where the city of Varanasi is located on high platform formed of pebbles and Kankars. Its religious and ritual significance is well established as it takes a westward turn near Barthara Kaalan in Chiraigaon development block and Paranapur in Cholapur development block known as Paschimvahini Ganga, where thousands of people take bathe on the day of Makar Shankranti. From Chandrawati to Kaithi (in Cholapur development block) it flows northward, resembling to its flow at Varanasi, thus Kaithi is known as Up-Kashi. The rivers Varuna and Gomati are the main tributaries of the Ganga (Rai and Mohan 2014).

2.4.2 *The Gomati River*

The River Gomati, a meandering river touches the Varanasi district near village Bhadwan in Cholapur development block and forming its northern boundary (33 km) emerges into the Ganga flowing through the village Niyardih, Babatpur, Azagara, Dhaurhara, Tekuri, Rajwari and Kaithi. Its main tributary is the river Nand which joins river Gomati at Ajgara village (Rai and Mohan 2014).

2.4.3 *The Varuna River*

The Varuna a perennial river, divides the Varanasi district into two part and joins the river Ganga near Sarai Mohan east of Varanasi city after flowing for nearly 40 km. vis Akohra, Kundi, Gaharwar, and Koirajpur villages. It has no important tributary excepting Bisuhil. Throughout its course, the river Varuna has a impartially high bank which is scoured on either side by many ravines (Rai and Mohan 2014).

2.4.4 *The Assi River*

Assi is a small local rain fed stream, the old historical southern boundary of Varanasi, highly polluted and congested, chocking due to encroachment. The Ganga, Varuna and Assi are the three natural streams which finally receive the storm water flows into the city of Varanasi. A number of interconnected ponds play a significant role in the drainage of the storm water one of the prominent rivers of our country (Rai and Mohan 2014).

The Nand is non-perennial tributary of the river Gomati. It originates from a low-lying land (Tal) of Karikhiyon in Pindra development block and flows towards the east, debouches in to the river Gomati after traversing a distance of 35 km. Through there are a number of unnamed nullahs, which join the river but they are dry throughout the year except in the rainy season. Hathi nala, a tributary to the river Nand, is disastrous sufficient in the rainy season but in the summer it is entirely dried up (Rai and Mohan 2014).

2.5 Climatic Characteristics of the Varanasi District

Climate is the most vital factors which determine the pattern of land utilization. It strongly influences the activities of man. The attributes of climate i.e. rainfall, temperature, pressure, wind direction and velocity and humidity have a good affinity with agricultural patterns. Since crop producing ability of an area depends largely upon climatic and soil condition, so the role of these elements cannot be overlooked for the purpose of ascertaining the agricultural activities.

The area under study lies in the sub-tropical monsoonal climate which is divided into the following three seasons:

(i) A cool dry season of northerly winds from October to February.
(ii) A hot dry season from march to mid-June and
(iii) A hot wet season of south-westerly wind from mid-June to September.

2.5.1 Temperature

From October onward till February the weather progressively crystallises into dry cool season with average temperature fluctuating from 15.46°C to 26.03°C when cold waves sweep over the area from west or north-west, nights becomes chilly and foggy sometimes giving frosty weather also. The western disturbances (Cyclones etc.) bring rainfall through small in quantity in this season which is very valuable to the Rabi crops.

By March, the day and nights go on becoming warmer in the region and heat continues to increase. From April, the weather quickly turns warmer and finally the hot and dry summer season sets in which is reflected in general by a sharp increase in temperature and relative humidity.

In the study area, temperature begins to fall rapidly from November and in January (The coldest month) the maximum temperature comes down to 23.06°. The average seasonal temperature varies from 24.65°C to 35.05°C but when the sun heat becomes oppressive followed by boiling sun rays and westerly hot winds (locally known as 'Loo'), mercury often to a maximum of 42.43°C. Generally the monsoon bursts in the study area during the third week of June. These hot winds naturally finish by mid-June with the arrival of south-west monsoon. Some times during this period in implication with cold waves, western disturbances carry the minimum temperature upto one or two degree above the freezing point normally liable for occurrence of frost.

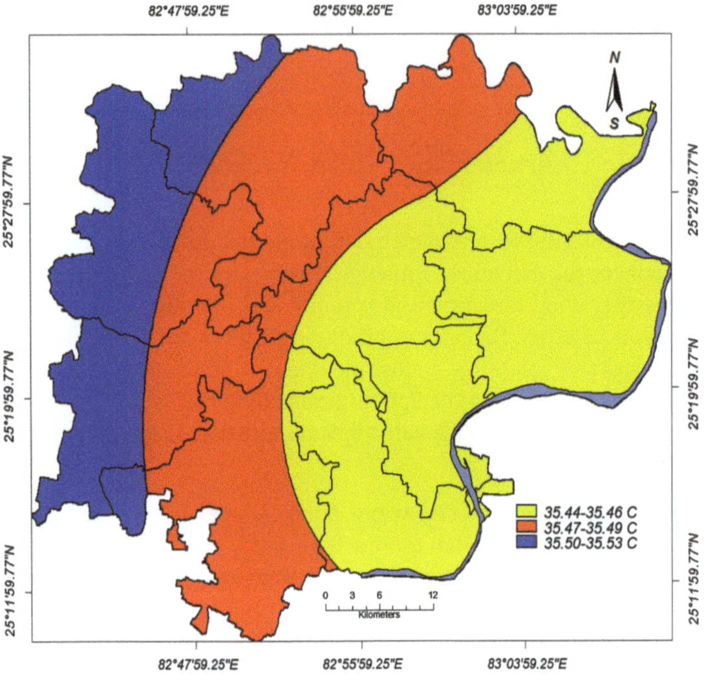

Fig. 2.5 Average distribution of temperature (in degree centigrade) in Varanasi district (1991–2010)

The hot wet season of south-westerly wind in the area starts from the third week of June and continued upto the end of September. Decrease in mean temperature upto a minimum of 27.20°C and maximum of 30.13°C, increase in relative humidity, high precipitation and low monthly range of temperature etc. are the important features of this season. Figure 2.5 shows the average temperature of Varanasi district of last 20 years.

2.5.2 Precipitation/Rainfall

The rainy season in the district usually instigates by the end of June and stays till the end of September. This is the utmost essential period for agriculture as it receives nearly 113.47 cm annual rainfall. The moisture content in the soil either by infiltration or capillary rise is fulfilled by rainfall and it is the major ecological factor in agricultural operations starting from cultivating to sowing, growth and maturing of all the crops. Most of the 'Kharif' crops depend on rainfall while occasional winter rainfall favours Rabi crops although it is mostly dependent on irrigation. Figure 2.6 shows the average rainfall condition of Varanasi district of last 20 years. Normally the monsoon starts in the study area from the third week of June and ends by the

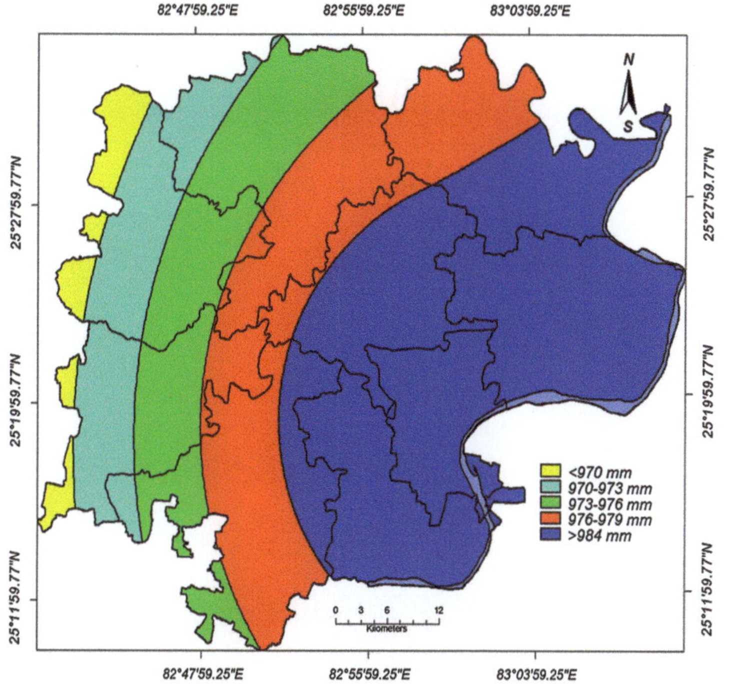

Fig. 2.6 Average distribution of rainfall (in mm) in Varanasi district (1991–2010)

September. Rainfall was consistently dispersed throughout the area. In 2009, the total annual rainfall is recorded as 102.5 cm of which about 90 % is attributed to monsoon periods (Table 2.1).

2.6 Ground Water: Present Status and Prospects

Varanasi is short of water. The current water supply system is not adequate to supply to around 1.6 million people in the city. There is a substantial gap between the recommended norm and the actual water supply. The city obtains a total of 270 million liters water from the river Ganga and ground water source.

The city, a part of central Ganga alluvial plain, is underlain by the quaternary alluvium containing fine to coarse grained sand, clay and clay with Kankar. The alluvium belongs to the quaternary group of pleistocene system of the recent geological age. To be specific, the older alluvium is middle to upper pleistocene and the newer alluvium is latest (Krishnan 1960).

Study of the upper 120 m of alluvium based on archives of tube wells in the area designates the presence of the two distinct sedimentary horizons, namely (i) backswamp clays comprising Kankar at places lying immediately below the land surface

Table 2.1 Climate data of Varanasi district, 2010

Month	Jan.	Feb.	Mar.	Apr.	May	Jun.	Jul.	Aug.	Sep.	Oct.	Nov.	Dec.	Year
Average high °C	19	24	31	37	38	36	32	31	31	31	27	22	29.9
Average low °C	8	12	17	22	25	27	26	26	24	21	15	11	19.4
Precipitation (mm)	19.3	13.5	10.4	5.4	9.0	100.0	320.6	260.4	231.6	38.3	12.9	4.0	1025.4

Source: B.H.U., Varanasi, 2010

and having an average thickness of about 50 m and (ii) the underlying meander-belt deposits containing of mixed populations ranging from fine to coarse sands, having an average thickness of around 60 m. In these two unconsolidated alluvial horizons, ground water arises in the pore space in the zone of saturation (Pathak 1977).

These two ground water bodies are hydraulically separated where the back swamp deposits are thick but are likely to be interconnected where they are comparatively thin. These two types considerably differ in their capacity to absorb water at the land surface and to transmit water to the zone of saturation. The back-swamp clays are usually dense and impervious and cover the entire area. These clays are enormous and comprise Kankar and thin lenses of fine-grained sands. Because this material is generally too fine grained to transmit water readily, it prevents the downward percolation of water from the land surface and also confines water in the underlying meander-belt sand deposits. These deposits constitute a ground water reservoir from which very large supplies of water can be recovered perennially for irrigation, municipal and industrial purposes. The shallow ground water in the back swamp deposits (clay and Kankar beds) is mostly unconfined and its static water level is only a few metres below the ground level. It supplies water to a good number of dug wells, bore wells and wells fitted with small pumps used both for domestic or irrigation purposes. The meander belt sand deposits from the main ground water body in the area and supply water to a large number of tube wells used for irrigation, industrial and domestic purposes. The deep ground water body, measured to be hydraulically continuous, is known to be confined locally. The piezometric surface of the deep body varies and is generally identified to be deeper in comparison to the static water level of shallow water body. It should be stated that the ground water table in the city has gone down to 4.6 m (15 ft) below the level it used to be prior (Mohan et al. 2011).

As the texture and thickness of the meander-belt deposits differ significantly, their water yielding capability also fluctuates from place to place. Most of the tube wells in the city originate water from meander-belt deposits. These tube wells yield (30–60) L s^{-1}water. A few tube wells which originate water from the thin disengaged sand lenses in back-swamp clay yield less than 15 L s^{-1} (Mohan et al. 2011).

In 1999, the Municipal Corporation (now Nagar Nigam) completed a survey and recognized 5000 wells in the city. However, present state-of-art is scruffy; it is identified that about 2200 wells are in a workable condition. Observably the rest of the wells were occupied with wastes, garbage's etc. (Mohan et al. 2011).

2.6.1 Current Status of Water Supply

State government department, Varanasi Jal Sansthan is affianced in the supply of drinking water to the people living in the city of Varanasi. This Jal Sansthan draws drinking water from two sources: (a) the river Ganga and (b) ground water.

Bhadaini Intake Works is pumping a total of 1.25×10^5 m^3 d^{-1} raw water of the river through its six pumps, four each with a capacity of 40 m^3 min^{-1} and two each with 30 m^3 min^{-1}. The pumped raw water is sent to the Bhelupur Water Works, situated at about 1.5 km away from the water drawing point, for its chlorination and purification. The treated water is supplied to the inhabitants of the city through a network of pipelines the total length of which is about 575 km. This water supply system was laid way back in 1892 by James Princep. Sometimes, because of failure of electricity supply the Bhadaini intake works fail to pump up water from the river to its full capacity and because of which the problem becomes acute. At limited places, quite often, public of the city increase a hue and cry about mix up of water with sewage because of leak in the pipes. The various gastro-enteric diseases and other health hazards support this type of claim (Mohan et al. 2011).

Along with the river water, a total of 1.45 m^3 d^{-1} of ground water is extracted from 118 deep bore tube wells installed in different localities of the city to supply to the growing demand of water. Further, a total of 2347 hand pumps were also set up at different parts of the city. Jal Sansthan of Varanasi has classified the city area into 16 zones (11 zones within the city spread between Ganga and Varuna and five zones in the trans–Varuna area (located to the other side of the river Varuna, as a western extension of the city) for the management and distribution of water supply. The trans-Varuna is dependent completely on ground water resource (Mohan et al. 2011).

With all types of water supply put together, Jal Sansthan delivers a total of 2.70×10^5 m^3 d^{-1} water. With such a volume of supply, each person in the city gets 169 L d^{-1} water, which is far less the WHO's norm of 270 L d^{-1}. Thus one fifth of the population of the city is not supplied with drinkable water (Mohan et al. 2011).

The river Ganga is an imperative source of drinking water in the city, but the level of water goes down for several reasons, below the intake points. Jal Sansthan finds it very tough to drink 1.25×10^5 m^3 d^{-1} water not only during spring and winter but also during some period of rainy season. Population growth in a geometric series could be measured responsible in the area (Mohan et al. 2011).

2.6.2 *Available Resources*

The city, a part of Central Ganga Alluvial Plain, is underlain by the Quaternary Alluvium comprising fine to coarse grained sand, clay and clay with Kankar. The alluvium belongs to the quaternary group of pleistocene system of the recent geological age. To be precise, the Older Alluvium is Middle to Upper Pleistocene and the Newer Alluvium is recent (Krishnan 1960; Mohan et al. 2011)

Study of the Upper 120 m of alluvium based on records of tube wells in the area specifies the presence of the two distinct sedimentary horizons, namely (i) back-swamp clays comprising Kankar at places lying instantly below the land surface and having an average thickness of approximately 50 m and (ii) the underlying meander-belt deposits containing of mixed populations ranging from fine to coarse sands, having an average thickness of about 60 m. In these two unconsolidated alluvial horizons, ground water occurs in the pore spaces in the zone of saturation (Pathak 1977; Mohan et al. 2011).

These two ground water bodies are hydraulically unglued where the back swamp deposits are thick but are possible to be interrelated where they are comparatively thin. These two types significantly vary in their capacity to absorb water at the land surface and to transmit water to the zone of saturation. The back-swamp clays are generally dense and impervious and cover the entire area. These clays are massive and contain Kankar and thin lenses of fine-grained sands. Because this material is normally too fine grained to transmit water readily, it prevents the downward percolation of water from the land surface and also restrains water in the underlying meander-belt sand deposits. These deposits constitute a ground water reservoir from which very large supplies of water can be recovered perennially for irrigation, municipal and industrial purposes. The shallow ground water in the back swamp deposits (clay and Kankar beds) is usually unconfined and its static water level is only a few metres below the ground level. It supplies water to a good number of dug wells, bore wells and wells fitted with small pumps used both for domestic or irrigation purposes. The meander belt sand deposits form the main ground water body in the area and supply water to a large number of tube wells used for irrigation, industrial and domestic purposes. The deep ground water body, considered to be hydraulically continuous, is known to be confined locally. The piezometric surface of the deep body varies and is generally known to be deeper in comparison to the static water level of shallow water body. It should be mentioned that the ground water table in the city has gone down to 4.6 m (15 ft) below the level it used to be earlier. The ground water level was at a general depth of 12 m–18 m (40–60 ft) in the year 2005; while in the year 2006 it was between 17 m and 21 m (55–70 ft) deep (Report 2006; Mohan et al. 2011)

As the texture and thickness of the meander-belt deposits differ significantly, their water yielding capacity also differs from place to place. Most of the tube wells in the city derive water from meander-belt deposits. These tube wells yield (30–60) L s^{-1}water. A few tube wells which derive water from the thin disconnected sand lenses in back-swamp clay yield less than 15 L s^{-1}.

In 1999, the Municipal Corporation (now Nagar Nigam) completed a survey and identified 5000 wells in the city. But, a recent survey described only 2204 wells in presence. Obviously the rest of the wells were occupied with waste, garbage etc. (Mohan et al. 2011).

2.6.3 Ponds as a Source of Recharge

The essential source to sustain ground water body, in fine to coarse-grained sands of the older alluvium, is rainfall that seeps down to the ground water table. Other sources contain infiltration from river, return seepage from agriculture irrigation. The annual average precipitation (50 years average) in the city is 1076 mm. In the year 2008, the city desired for rains and witnessed a rainfall of only 300 mm. There is no recognized system of ground water recharging in the city i.e. developments of water bodies, ponds and tanks etc. Most of the rainwater goes wastes and join the river Ganga ultimately (Mohan et al. 2011).

In Varanasi, there used to be many ponds and tanks dating back to ancient time. Also serving as the holy places for holding Hindu religious rituals, they also played a vital role in rainwater collection and thereby helped as sources for ground water renewal. Total 118 ponds and tanks in the city have been documented (Fig. 2.7).

However, due to rapid development of the city, most of these sacred ponds have been wiped out from the map of the city. Our investigation has shown that 44 ponds were substituted by settlements. This shows complete ignore of environmental conservation to accommodate growing population. Public unawareness of the importance of these ponds inferior the situation by completely or somewhat filling up many of the existing ponds with garbage and waste (Mohan et al. 2011).

Providentially at the request of avid citizens and environmentalists, the city authorities have arranged a list of 54 ponds and tanks which in need of urgent rehabilitation; the plan has not been efficiently applied though. These 54 ponds may soon die out if instant rehabilitation measures are not executed. Only about half a dozen of them have water all over the year. The Central Government has shown deliberation in the very direction and released a sufficient fund for the conservation and beautification of water bodies of Varanasi (Mohan et al. 2011).

2.6.4 Pollution

Pollution of the ground water in the Varanasi city is generally sourced from loads of garbage. The solid and liquid wastes generated out of the household and industrial actions are dumped and unconfined in uncontrolled sites. On an average, 661 million tons of solid wastes in a day are made in the city, but only 87 % of which is collected for ultimate disposal, and the rest is left uncollected. This is mostly due to lack of effective labor strength and fleet of vehicles for collection, transportation and disposal of solid wastes. These wastes are disposed of in the low lying areas of the city where the tanks, ponds and Ganga river are situated, which were once main sources of ground water recharge in the city. In order to know the consequence of wastes on ground water quality, water samples from 6 sites (hand pumps) around garbage dumps were studied. At two sites (Aurangabad and Badi Gaibi), samples were collected in two different (summer and rainy) seasons and the outcomes are recorded in Table 2.2. The ground water from these two sites appears to be unexceptionally affected by the solid waste dump in terms of chemical contents. Nevertheless, the turbidity is far above acceptable limit, slightly close to the maximum permissible limit. The hardness exceeds a value higher than double the adequate limit but still below the permissible limit. Unpredictably, the water is free of coliform (Mohan et al. 2011).

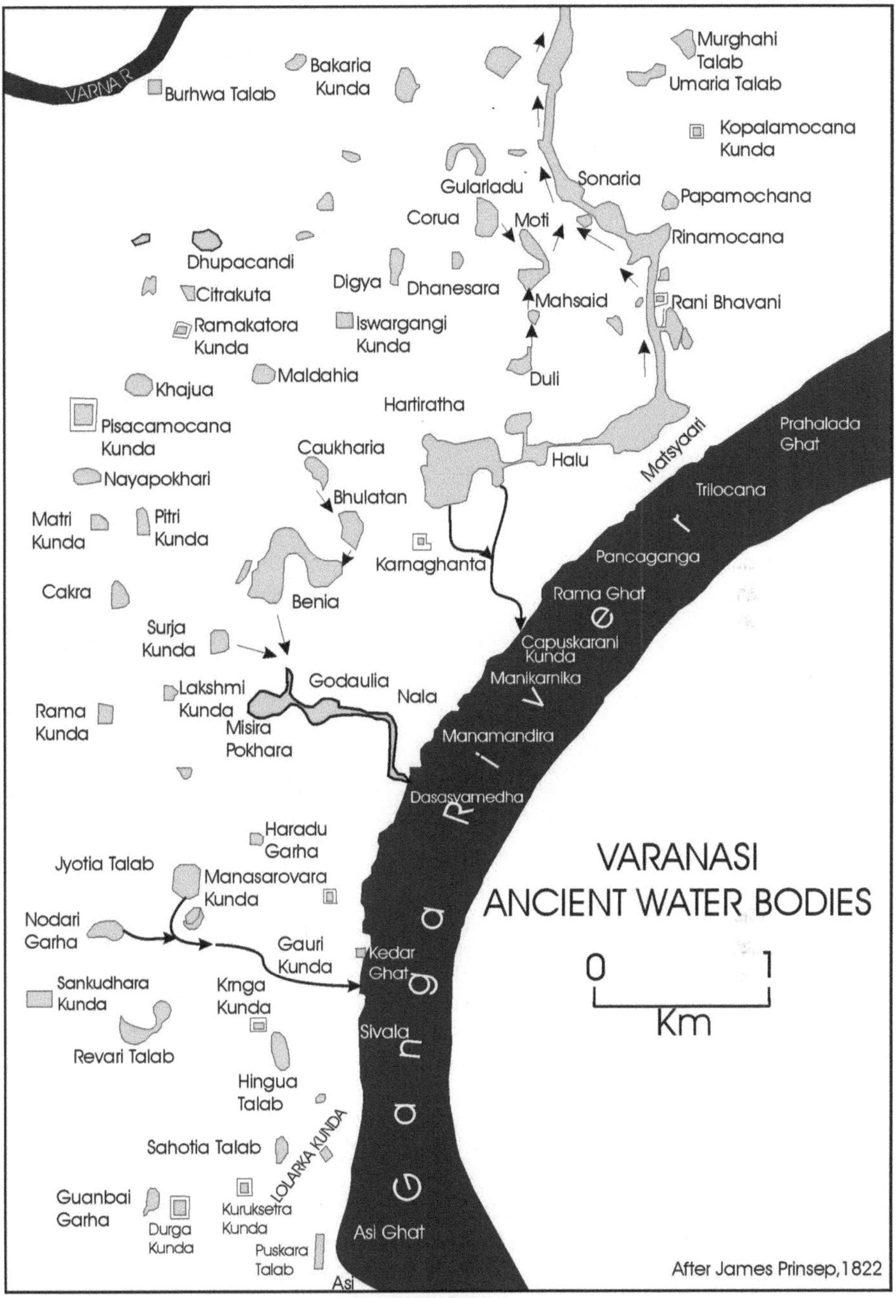

Fig. 2.7 Pattern of ancient water bodies in Varanasi, 1822 (Mohan et al. 2011)

Table 2.2 Quality of ground water in summer and rainy seasons, 2009

Parameters	Value				Acceptable limits (CPWB, N. Delhi[a])	Maximum Permissible limits (CPWB, N. Delhi[a])
	Site 1		Site 2			
	Summer season	Rainy season	Summer season	Rainy season		
1. Alkalinity	188	182	192	180	200	600
2. pH	7.94	7.02	7.03	7.17	6.5–8.5	9.2
3. Total dissolved solid	356	342	416	387	500	1500
4. Conductivity (umhos)	1.2	10	1.6	10	–	–
5. Hardness	438	408	406	428	200	600
6. Chloride	61	52	55	46	200	1000
7. Sulphate	16	16	18	12	200	400
8. BOD Biologic oxygen demand	0.2	0.4	0.6	0.4	Less than 4	
9. COD Chemical oxygen demand	1.2	4.0	3.6	4.4	–	–
10. Total suspended solid	15	35	19	29	100	–
11. Total coliform (MPN/100 ml)	NT	N.T	NT	N.T.	10	10
12. Turbidity (NTU)	10	05	10	02	2.5	10
13. Fae cal coliform (MPN/100 ml)	NT	N.T.	NT	N.T.	Nil	–
14. Nitrate	0.01	0.8	0.04	0.8	4.5	45

Note:
(i) All the results are expressed in mg/l except pH
(ii) *N.T.* not traceable, *NTU* nephlometric turbidity units
(iii) The test has been done in the laboratory of U.P. Pollution Control Board, Varanasi
(iv) The two sites selected were rather in the midst of waste dumps and its water are under use by the inhabitants for domestic purposes
(v) Site1: Aurangabad, Site 2: Badi Gaibi
(vi) [a]CPCB- Central Pollution Control Board, New Delhi. ADSORBS/3/1978–79 (Mohan et al. 2011)

The ground water in rainy season has to some extent lower alkalinity, TDS, turbidity, chloride and conductivity but higher TSS, COD, and nitrate. The water from Aurangabad has higher TDS, hardness, COD, and pH and that from Badi Gaibi has higher alkalinity, chloride, sulphate, TSS and turbidity. These spatial differences might be the consequence of seepage from different solid waste disposal (Mohan et al. 2011).

During the last rainy season (July-August) water samples were taken from hand pumps surrounded by waste dumps on 4 sites keeping in view the likely impact of

Table 2.3 Quality of ground water from site 3 Nakhighat, site 4 Benia Bagh, site 5 Pichasmochan and site 6 Nagwa police Chowki, 2009

Parameters	Value				Acceptable limits (CPWB, N. Delhi[a])	Maximum Permissible limits (CPWB, N. Delhi[a])
	Site 3	Site 4	Site5	Site 6		
1. pH	7.5	7.2	7.5	7.1	6.5–8.5	9.2
2. TDS	316	648	624	600	500	1500
3. Hardness	172	152	160	164	200	600
4. Chloride	16	168	168	164	200	1000
5. Sulphate	5.0	90	50	Nil	200	400
6. Floride	Nil	Nil	0.8	Nil	–	–
7. NO_3 (Nitrate)	5.0	20	25	Nil	4.5	45
8. NO_2 (Nitrite)	Nil	0.01	0.15	Nil	0.1	–
9. Calcium	26	29	30	24	–	–
10. Magnesium	26	20	21	26	–	–
11. Iron	0.3	0.4	0.3	1.0	–	–
12. Colour	<5	<5	<5	10	–	–
13. Taste/ Odour	Unobjectionable	Brackish	Brackish	Brackish	–	–
14. Turbidity	Nil	1.2	3.6	12	2.5	10

Note:
(i) All the results are expressed in mg/L except stated otherwise
(ii) The test has been done in laboratory of U.P. Jal Nigam, Varanasi
(iii) The sites selected are in the midst of waste dump and its water is under use by the inhabitants for domestic purposes
(iv) [a]CPCB-Central Pollution Control Board, New Delhi. ADSORBS/3/1978–79 (Mohan et al. 2011)

the dumps on the ground water quality. The parameters measured were again 14; the investigation was completed in the laboratory of the U.P. Jal Nigam. The outcomes acquired are shown in the Table 2.3. TDS is well above the standard limit though within permissible limit in ground water from 3 of the 4 sites premeditated. The color (except at Nagwa Police Chowki), taste/odour (brackish) and pH are more or less normal. So are the calcium, magnesium and iron contents. Among the 4 sites, iron is noticeable at Nagwa Police Chowki site (Mohan et al. 2011). Here, the water is considered by highest quantity of turbidity (12 NTU) in all 6 sites but is devoid of fluoride and nitrate. Ground water from Aurangabad and Badi Gaibi has higher chloride level than the water from the other 4 sites. Sulphate is maximum at Beniabagh. In the final investigation, only conspicuous result is the presence of NO_2, which is beyond the acceptable limit at Pisachmochan site. No doubt, there is a marked difference in the chemical parameters of water at different sites. In fact, the impact of solid waste dump, in general, is not beyond permissible limit which may be attributed to the deep (more than 150 ft) boring of hand pumps (Mohan et al. 2011).

In the Ganga Action Plan Phase-I, the construction of municipal sewage treatment plants in various segments of the city area i.e. Konia-Dinapur (in the northern sector) and Bhagwanpur (in the southern sector) was taken up. The treated sewage water, being plant nutrient, is extensively applied to irrigate crops in the area within surrounding area of treatment plants. Insistent leaching of the dissolved nitrate content of the liquid wastes downwards through the permeable irrigated soil to the top saturated aquifer horizon possibly induced enrichment of nitrate content in ground water of the city. Moreover, poor sewerage and drainage facilities, seepage of human excreta from very old septic tanks, and use of nitrogenous fertilizer might have also contributed to nitrate enrichment in the ground water (Mohan et al. 2011).

In total, pollution of ground water in the city is usually from urban and industrial wastes. However, the use of chemical fertilizers and pesticides in agricultural fields in the nearby sub-urban sectors for cultivation purposes has also increased the problem. A large variation of chemicals is being used by farmers in the crop fields to eliminate various types of crop pests, including insects and other organism. These pesticides find their way to ground water through percolation. Research on pollution of ground water in other parts of India has been stated copiously (Roychowdhury et al. 1999; Ahmed et al. 2001), but that of Varanasi is unusual (Rana et al. 2003; Singh et al. 2006). Varanasi is popular in the Indo-Gangetic plain for the production of vegetables and fruits. Thus, there was a rather long history of use of fat accumulative, highly toxic banned cyclodiene organo-chlorine pesticides in the city for agricultural farming purposes and also to retain public hygiene (Rana et al. 2003; Singh et al. 2006).

The higher amounts of aldrine in the groundwater in rural areas appear to be related to its extensive use in plant protection of cereals, vegetables and fruits for the last few decades. The relative concentrations of cyclodiene organo-chlorine pesticides in ground water were in the following order: aldrine > chlordane > dieldrine > heptachlore and its epoxide. The study (Rana et al. 2003; Singh et al. 2006) specified that the banned cyclodiene organo-chlorine pesticides so far analyzed both in the rural and urban areas have crossed the FAO/WHO (World Health Organization) limit of water quality and contamination of these cyclodiene pesticides is in alarming stage for human and livestock consumption and the possibility of their accumulation in food chain cannot be ignored (Mohan et al. 2011).

The higher amount of aldrine in groundwater in urban area seemed to be associated to its widespread use in house hold purpose and plant protection of vegetables in kitchen gardens as well as public health programs through local municipality (Rana et al. 2003; Singh et al. 2006). UNICEF described the presence of 'arsenic in the ground water of southern parts of the city at a level up to 499 ppb far beyond the limit (10 ppb) set by the WHO and Bureau of Indian Standard for the potable water. The joint investigation committee comprising the representatives of Chief Medical Officer, Nagar Nigam and Jal Sansthan analyzed the ground water in the city and claimed that 75 % of the water supplied through tube wells was safe and the rest was unsuitable for drinking purpose, through a orthotolidine test on quantity of chlorine in the water. But, the National Institute of Communicable Disease (NICD) has rejected this type of test by stating that the test lacks scientific basis. Nagar Nigam does not carry out coliform test to measure E. coli organism. The ground water in the first and second strata is polluted from nitrate and faecal coliform organisms. Banaras Hindu University in the southern part of the city fulfills its demand of water from the third strata occurring at 200 m below ground level through deep bores (Mohan et al. 2011).

The Central Government has initiated a plan in 1986 namely ‘Rajiv Gandhi National Drinking Water Mission’ to provide safe drinking water. The Mission has been entrusted to analyze the samples of ground water with respect to the levels of iron, fluoride, nitrate and total dissolved solid (TDS). Unfortunately, under this mission not much could be implemented in the State of Uttar Pradesh due to apathy of the State Government.

Although crisis of water is prevalent in the whole city, the southern part is suffering more owing to lowering of ground water each year. Brij Enclave colony in Varanasi is a densely populated area of the southern part of the city.

Here, ground water table dropped by 2.13 m (7 ft) in 2006 from a level of 17.68 m (58 ft) in 2005 (Mohan et al. 2011).

Report of the State Ground water Department states that the ground water in the city of Varanasi is depleting at a rate of 23 cm /a^{-1} (Report 2006). Lowering of ground water in the southern part is nine times faster than the rest part of the city. The Central Government has recently asked State Governments to prepare an integrated development plan for a well-planned township under the proposed ‘Jawahar Lal Nehru Urban Renewal Mission’, so that it might meet the demand of rapidly growing population for the next 25 years. The State Government has submitted its detailed feasibility report to the Central Government for integrated development of the proposed ‘Greater Varanasi’, which includes the southern part of Varanasi city also. Therefore, the southern part, the most affected part in terms of water supply, might be developed as a township under the Mission (Mohan et al. 2011).

2.7 Soils

Soils are the most valuable resources for the luxuriant growth of natural vegetation and agricultural plants. Crop plants, to a great extent depend upon the good quality of soil. Generally, soils in the study area are part of alluvial soil. Soils in the study area are extracted through IRS-1C LISS III data of 2008 by applying image processing supervised classification method (Fig. 2.8). The soils of Varanasi district can be divided into four types depending upon their texture composition and formation process as the rivers have played a major role in it:

(i) Ganga Sandy Loam
(ii) Western Low Land Soil
(iii) Western Upland Soil
(iv) Loamy Soil.

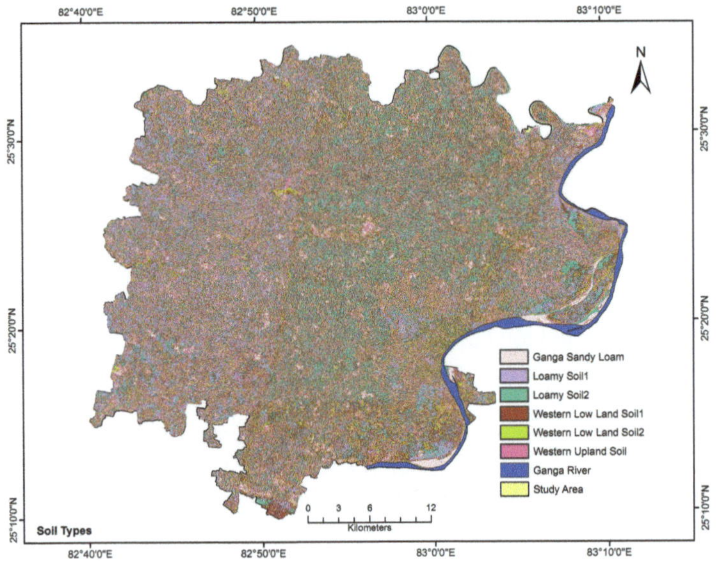

Fig. 2.8 Soil in Varanasi district prepared through IRS-1C LISS III satellite data, 2008

2.7.1 *Ganga Sandy Soil*

It covers 25 % of the area along the bank of the river Ganga and lower part of the river Gomati especially lying in the inundated or flooded areas in the rainy season. Sandy soil has either sands or a mixture of small fine particles of soil suitable for Zaid crops.

2.7.2 *Western Low Land Soil*

It occupies the northern part of the district along the Gomati river and its tributary. It is a fertile loamy soil and neutral in character varying from brown in colour with sufficiently mature profile. There are certain usar and semi usar (barren land) patches scattered here and there in Cholapur and Chiraigaon development blocks. They are unclassified and lack in well development profiles. These are moderately alkaline in nature with a pH value of 8.2.

2.7.3 *Western Upland Soil*

It is a stiff clay loam type of soil and covers most of the areas of the district intermixed with alluvial soil in different parts. In some scattered patches this soil is affected by saline or salt affected locally called as Reh or salt affected soil. In these patches this patches, hard formation can be seen at a depth of about 5 m below the ground which restricts infiltration of rain water (Rai and Mohan 2014).

2.7.4 Loamy Soil

This soil is found in the northern part of the Cholapur development block along the bank of river Gomati. The soil consists of fine and coarse sand. It is deficient in plant nutrients with low water holding capacity.

In general, the soils of the district may broadly be divided in to two major types:

(i) Khadar and
(ii) Bhanger

A Khadar (newer alluvium) covers the flood plains in the vicinity of the rivers including their reaches and also old beds. It is replenished annually by new deposits. The soil naturally remain moist as its derives moisture from the river through seepage in dry season and are capable of growing Rabi and Zaid crops without irrigation. These are locally known as Tari and are classified as Ganga sandy soils consisting of fine silts along the rivers Ganga and Gomati. The newer alluvial soil contains low percentage of humus content and nitrogen. Being more silty in texture it is highly friable and favorable for the Bhadai (Kharif), Rabi and Zaid crops (especially root crops). Under secured irrigation facilities, some of the areas with this soils are producing rice too, especially the early varieties (Rai and Mohan 2014).

The Bhangar is older alluvium with no flood effect except the low lying areas in patches which are either termed as Tal or consist of impervious clay layer found in the sub soil. It is often heavier with higher proportion of clay. Being sticky and usually not well drained it is richer in lime content and kankar and is more suitable for rice cultivation. The low lying areas which hold water for a long time and become dry in winter season is good for cultivation of Rabi crops (Rai and Mohan 2014).

2.8 Land Use and Natural Vegetation

For a long time human existence was based on plough and pastoral activities. It has almost destroyed all the forest covers lying earlier in the study region by direct cutting, burning and grazing. Presently the area under forest is almost negligible. The total area under forest was only 22 ha in year 2001. Only in the Tari region of the district it is observed in its natural form. In cultivated areas it is found in the form of gardens, groves and trees planted under social forestry. Mango and Banyan tress are observed throughout the region. Transplantation of trees under social forestry programme has started in the study area along the railway lines, canals and roads on vacant lands of Gram Panchayats. Munj, sarpat and some local grasses and bushes along the rivers and undulating lands provide a scenario of the natural vegetation through their natural panorama is distributed and altered now days by the inhabitants to fulfill their growing demand for building materials and other domestic users (Rai and Mohan 2014).

With growing population and limited land resources the significance of land use planning is obvious. Land has limited carrying capacity beyond which there will be degradation and forfeiture in yield due to too much use (Rai and Mohan 2014). In order to meet various demands of the growing population the land degrading trend needs to be checked.

The revenue department divides land uses in following categories: (i) Land place to non-agricultural uses, (ii) Barren and uncultivable land, (iii) Pastures and grazing land, (iv) Land under trees and orchards, (v) culturable waste land, (vi) current fallow, (vii) Fallow other than current fallow, (viii) Net Sown area, (ix) Forest.

Using IRS-1C LISS III remote sensing data of 2008, land use land cover (LULC) map is prepared using image processing supervised classification technique, in which five important classes are delineated i.e. Agricultural field, Fallow land, Vegetation, Built-up Area, and Water Bodies (Fig. 2.9). The Study area mainly comes under the Ganga plain having high fertile soils, so that maximum area comes under very good agricultural land. Forested area is being reduced by pushing the frontier of agriculture. On the other side good agricultural land is being grabbed by urban sprawls, industrial expansion and expansion of urban settlements and infrastructural development. Table 2.4 shows the area statistics of important land use

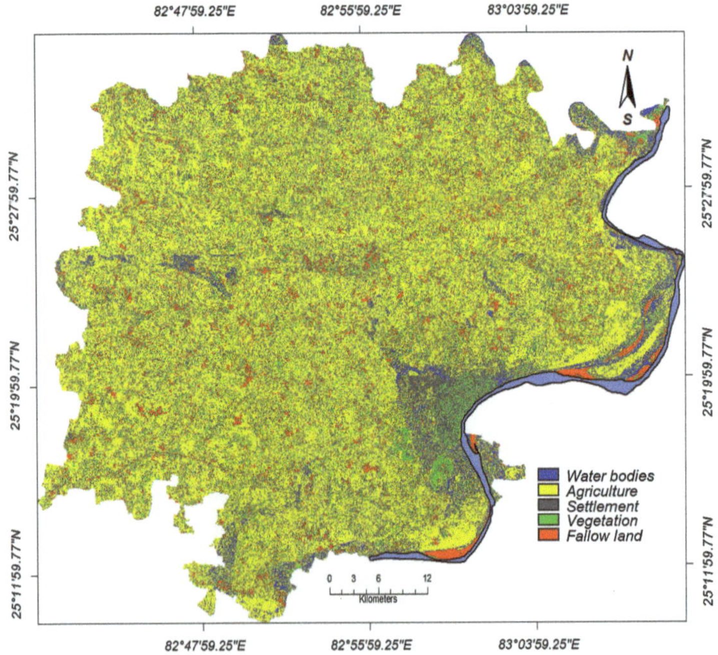

Fig. 2.9 Land use/land cover of Varanasi district, 2008

Table 2.4 Area statistic (in hectares) of different land use categories in Varanasi district

Year /Block	Barren cult. waste	Present fallow land	Other fallow land	Barren land	Land put to non – Agri. use	Pastures	Area under bush, forest & garden
2006–2007	3014	20,846	2264	2150	25,556	10	2996
2007–2008	2932	20,945	2314	2151	25,600	24	2964
2008–2009	2153	22,182	2410	2247	26,097	23	2961
Blockwise, 2008							
Baragaon	183	2464	112	299	2400	3	258
Pindra	234	2546	160	199	2713	0	281
Cholapur	104	2705	212	153	2137	2	311
Chirai Gaon	534	2480	225	519	3439	1	266
Harhua	205	1740	207	146	1973	2	221
Sevapuri	290	1837	152	155	2430	13	544
Arajiline	181	2474	243	159	2880	0	459
Kashi Vidya Peeth	140	4776	483	273	3935	2	479
Total rural	1871	21,022	1794	1903	21,907	23	2819
Total urban	282	1160	616	344	4190	0	142
Total district	2153	22,182	2410	2247	26,097	23	2961

Source: Statistical Handbook, Varanasi District, 2009

features in the Varanasi district. The district covered by 2961 ha forest and dense bushes. However, it is quite below than the standards of 'Natural Forest Policy' which said that 33 % of land should be covered with forest so that greenery may be conserved.

2.9 Transportation Network and Connectivity

Varanasi is very well interconnected by road network, rail and air with rest of the country. Important cities like New Delhi, Lucknow and Allahabad are approximately 750 km, 260 km. and 125 km. away from here. There are three national highways i.e. NH-2, NH-56 and NH-29 and four state highways i.e. SH-87, SH-73, SH-74 and SH-98 passing through the city. The network linkages provided by the National highways are:

(i) NH 2- G.T. Road from Mughal Sarai to Allahabad;
(ii) NH 29- Varanasi to Gorakhpur, Kushinagar; and
(iii) NH 56- Varanasi to Lucknow via Jaunpur.

These national highways and state highways have high passenger traffic as these roads deliver a good connectivity to the adjacent areas in the U.P. state as well as to metropolitan cities like Delhi and Kolkata. The Grand Trunk (GT) road or NH2

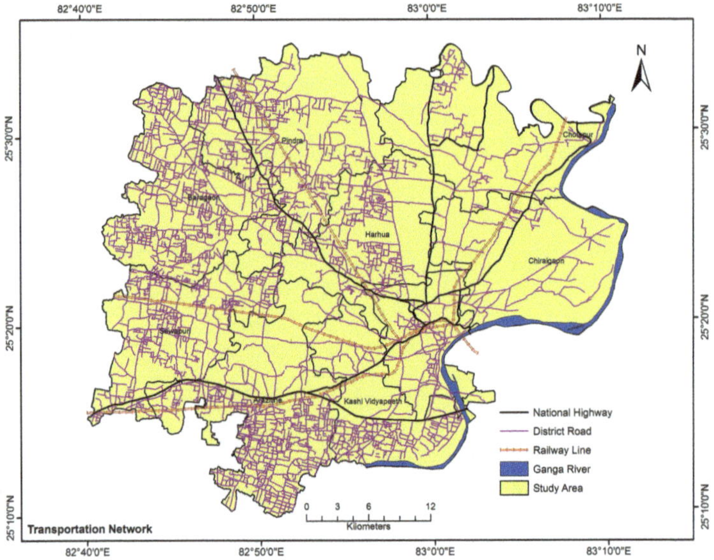

Fig. 2.10 Transportation network in Varanasi district

forms the main transportation backbone of the city. A bypass is being built along the Eastern edge of the city to release the burden off NH 2. Another ring road is constructed along the western edge of the city to divert the traffic and deliver better connectivity to the recent developments coming up in the Trans Varuna Region.

Varanasi is well joined by railway lines. There are three rail lines entering to the city from Lucknow, Bhadoi and Allahabad and is diverted in two lines to Gorakhpur and Mughal Sarai. The city lies on Delhi-Kolkata rail route of North Eastern Railways, which is the broad gauge. A rail line connects the town with Sarnath. The other cities having good connectivity through railways are Patna, Guwahati, Chennai, Mumbai, Gwalior, Meerut, Lucknow, Kanpur and Allahabad. The town also has an airport at a distance of about 24 km away from the city. Flights facilities from Varanasi to Agra, Bhubaneshwar, Kolkata, Delhi, Gorakhpur, Khajuraho, Lucknow, Raipur and Kathmandu (Nepal) are available. It is on a consistent aviation route of Delhi to Kolkata and Bhuvaneshwar. It is also the flying gateway to Nepal. Transportation network of Varanasi district has been extracted from IRS-1C LISS III satellite data in conjunction with IKONOS satellite data. Figure 2.10 shows the transportation network of Varanasi district. Table 2.5 shows that block-wise distribution of road network on the basis of rural population. Total 2610 km and 2615 km of metalled road are calculated in the year 2003–2004 and 2008–2009 respectively in which 1701 km of metalled road belongs to P.D.W. (public development work) in 2008–2009. In the rural and urban area of Varanasi, total length of metalled road is 1999 km and 616 km respectively.

Table 2.5 Block-wise distribution of road network on the basis of village population

Year/Development block		Length of metalled road (km)		No. of villages connected with all seasonable road (Population Wise)		
		Total	P.W.D	Village of <1000 population	Village of 1000–1499	Villages of>1500
2003–2004		2610	1696	433	169	267
2008–2009		2615	1701	433	169	267
Development block-wise, 2008–2009						
1	Baragaon	243	216	57	17	34
2	Pindra	240	178	55	25	34
3	Cholapur	219	155	44	21	26
4	Chirai Gaon	265	196	39	19	40
5	Harhua	233	172	67	27	25
6	Sewapuri	203	172	58	16	24
7	Araziline	289	237	64	18	43
8	Kashi Vidhyapith	307	253	49	26	41
Total Rural		1999	1579	433	169	267
Total Urban		616	122	0	0	0
Total District		2615	1701	433	169	267

Source: Statistical Handbook, Varanasi District, 2009

2.10 Communication and Educational Facilities

Development block wise distribution of communication facilities are shown in the Table 2.6 As for communication means concern, post office, PCO telephone etc. are comes under this category. In the Varanasi district, a total of 145 post offices, 1737 PCO, 10,941 telephone connections are available in the rural area whereas 135 post offices, 3088 PCO, 81,824 telephone connections are present in the in urban area. Total railway station/hault is predictable in the rural and urban area of Varanasi district is 15 and 06 respectively. Number of bus station in rural and urban area is 167 and 06 respectively.

Total 145 post office are located in the villages area of each development blocks, 296 post offices are located at the distance of <1 km, whereas only 58 post office are located at the distance of >5 km from the villages. Likewise, total 750 PCO are found in the villages whereas 820 PCO are identified at the distance of >5 km (Table 2.6). Distance wise different communication means in each development block are shown in the Table 2.7.

Educational facilities in the rural part of Varanasi district are not very good. Not any single university is established in rural development blocks but educational facilities in the city area are very good. Three important universities i.e. Banaras Hindu University, Mahatma Gandhi Kashi Vidyapith and Sampurna Nand Sanskti University are located in the Varanasi city. Development block wise status of educational facilities on per lakh population is shown in the Table 2.8. It was found that maximum 13 primary school on per lakh population were found on Kashi Vidyapith

Table 2.6 Block-wise distributions of communication facilities in Varanasi district

Year/Development block		Post office	Telegram office	PCO	Telephone	Railway station/ hault	Bus station
2003–2004		280	26	4294	94,140	21	173
2008–2009		280	26	4825	92,765	21	173
Development block-wise, 2008–2009							
1	Baragaon	20	3	148	1626	0	50
2	Pindra	23	1	184	1487	3	9
3	Cholapur	18	2	301	1542	2	17
4	Chirai Gaon	20	0	217	1366	0	12
5	Harhua	18	0	192	1322	1	8
6	Sewapuri	16	1	240	962	2	28
7	Araziline	18	3	302	1185	4	34
8	Kashi Vidhyapith	12	2	153	1451	3	9
Total rural		145	12	1737	10,941	15	167
Total urban		135	14	3088	81,824	6	6
Total district		280	26	4825	92,765	21	173

Source: Statistical handbook, Varanasi district, 2009

block whereas minimum 4.2 primary school on per lakh population were found on Baragaon development block in year 2008–2009 but total 3.2 higher secondary school and 13 middle school have been identified on Baragaon development and 28.1 and 13 higher secondary and middle school on per lakh population were found for Kashi Vidyapith block respectively. District as a whole, 9.7 primary, 29.3 higher secondary and 9.7 middle schools on per lakh population have been identified in the year 2008–2009 respectively.

2.11 Solid Waste Generation and Management

Any waste that is not in the form of gaseous or liquid state, such waste is known as solid waste. Any solid material that is disallowed by society is called solid waste. Solid wastes arise from human and animal activities and are normally discarded as useless or unwanted. In other words, solid wastes may be well-defined as the organic and inorganic waste materials formed by various activities of the society and which have lost their value to the first user. The quantity of municipal solid waste has increased enormously with improved lifestyle and social status of the populations in urban area (Sharholy et al. 2007). The anaerobic decomposition by microorganisms brings about degradation of most of the solid waste.

This results in the emission of carbon dioxide (CO_2), Methane (CH_4), and other toxic gases (Suchitra 2007 and Mc Bean et al. 1995). These air pollutants cause a plethora of health problems to the nearby residents (Shah 2007). During land filling of solid waste, continuous pressure results in the quizzing of a contaminated liquid

Table 2.7 Distance-wise distributions of different communication means in each development blocks

	In Village				< 1 km				1–5 km				> 5 km			
Development block	Post office	Letter box	Telegram office	PCO	Post office	Letter box	Telegram office	PCO	Post office	Letter box	Telegram office	PCO	Post office	Letter Box	Telegram Office	PCO
Baragaon	20	20	3	67	32	32	3	6	80	79	57	40	6	5	6	84
Pindra	23	23	1	79	76	76	11	12	73	73	13	54	27	17	17	164
Cholapur	18	18	2	108	13	13	3	4	109	109	37	25	2	3	3	101
Chirai Gaon	20	21	0	86	11	11	0	5	104	103	14	25	3	0	0	121
Harhua	18	18	0	96	62	62	4	9	82	82	67	36	0	6	6	97
Sewapuri	16	16	1	100	10	10	1	6	113	113	49	38	15	42	42	130
Araziline	18	18	3	130	86	86	19	6	108	108	101	37	4	5	5	94
Kashi Vidhyapith	12	12	2	84	6	6	3	14	75	75	85	18	1	26	26	29
Total	145	146	12	750	296	296	44	62	744	742	423	273	58	104	105	820

Source: Statistical Handbook, Varanasi District, 2009

Table 2.8 Block-wise distribution of status of educational facilities on per lakh population

S.No.	Development block	No. of primary school on per lakh population			No. of higher secondary school on per lakh population			No. of middle school on per lakh population		
		1990–1991	2002–2003	2008–2009	1990–1991	2002–2003	2008–2009	1990–1991	2002–2003	2008–2009
1	Baragaon	3.4	5.7	4.2	2.7	4.9	3.1	5.6	13	13
2	Pindra	3.1	6.6	6.6	12.2	23.5	26.5	3.1	6.6	6.6
3	Cholapur	4.9	6.8	6.8	15.4	27.7	32.1	4.9	6.8	6.8
4	Chirai Gaon	2.7	9.7	9.7	11.9	20.5	32.3	2.7	9.7	9.7
5	Harhua	4.2	9.6	9.6	12	20.4	32.4	4.2	9.6	9.6
6	Sewapuri	5.7	10.1	10.1	14.5	26.5	24	5.7	10.1	10.1
7	Araziline	3.4	9.0	9	11.1	21.7	28.6	3.4	9.0	9.0
8	Kashi Vidhyapith	2.3	13.0	13	14.0	21.1	28.1	2.3	13.0	13.0
Total		3.4	9.7	9.7	12.6	22.5	29.3	3.1	9.7	9.7

Source: Statistical Handbook, Varanasi District, 2009

Table 2.9 Annual trend of solid waste generation in the Varanasi city

Year	Solid waste generated (Tones)
1991	165,375
1994	172,100
1997	182,500
2002	200,750
2008	219,000

Source: Nagar Nigam, Varanasi, 2009

as leachate (Suchitra 2007). Leachate is a liquid emanating from a solid waste land disposal site that contains dissolved, suspended and microbial contaminants. The leachate has high organic contents, soluble salts and other constituents capable of polluting groundwater. This polluted water is unfit for drinking purpose and causes jaundice, nausea, asthma, miscarriage and infertility.

In Varanasi district there are 85 Ganga ghats with typical structure on the river Ganga which are frequented by a large number of pilgrim's everyday for taking a holy dip in river Ganga. About 100 narrow lanes not wider than 1.8 m having houses and commercial establishment on both sides leading to 85 ghats. The city along the Ghats has facing the severe serious problems of solid waste management due to congested lanes. Table 2.9 shows the total annual generation of solid waste during the period 1988–2007.

At present total quantity of wastes in Varanasi city is about 650 metric tones (MT) per day (Table 2.10) of which largest amount of about 425 MT solid waste generated by household or residential areas followed by Slaughter houses, construction waste, hospital and dispensaries waste and slum waste and solid waste generated daily by shops and vegetable/ fruit markets and worship places (temples etc.).

Table 2.10 Waste generated by different actions in Varanasi city

Type of waste	Quantity of waste generate (Metric Tons)
Household	425
Shops and offices	20
Hotels, lodges, guesthouses	15
Vegetables/Fruit Markets	20
Fish/ Meat Market	5
Place of worships	20
Construction waste	25
Hospitals and dispensaries	25
Slum Waste	25
Industries, workshops, work sheds	15
Slaughter houses	35
Beef Shops	10
Total	650

Source: Nagar Nigam, Varanasi, 2009

Table 2.11 Collection sites of solid waste in the Varanasi city

1.	Vidyapeeth	14.	Nadesar	27.	Khatikana
2.	Cantt.Railway Station	15.	Andhrapul	28.	Shivpur Hospital
3.	Circuit House	16.	Jaitpura	29.	Bakkrabad
4.	Kamalgarha	17.	Azadpark	30.	Adamkuan
5.	Golgadda	18.	Bhadaun	31.	Macchodari
6.	Kabirchura	19.	Chowk	32.	Sonia
7.	Harthirath	20.	Kotwali	33.	Kunjigarh
8.	Benia	21.	Harai Sarai	34.	Godaulia
9.	Prachi	22.	Rewari Dashashwamedh	35.	Nagar Nigam
10.	Aurangabad	23.	Fatman	36.	Rewari
11.	Akashwani	24.	Bhelupur	37.	Shivala
12.	Durgakund	25.	Kabir Nagar	38.	Pitarkunda
13.	Harishchandra	26.	Senpura	39.	Kaligarh

Source: Nagar Nigam, Varanasi, 2009

Out of which 450 metric tones (collection efficiency is 69 %) is collected daily in an unscientific manner. And the average generation rate of municipal solid waste is 0.59 kg/capita/day. Rapid growing population and change in life styles of the people are amicable factors in waste generation.

Varanasi city has a total number of 39 secondary waste collection points (Table 2.11 and Fig. 2.11). These solid waste collection sites have been located through GPS (Global Positioning System) in the year 2010.

The total number of open dumps either abandoned or in use is 74. All these open dumps are within the city limits. Some of the 39 sites are clubbed along with the

Fig. 2.11 Selected solid waste collection sites in Varanasi city identified through GPS, 2010

nearly sites to reduce the number to 28. Solid waste is being dumped on private lands e.g. all open dumps in Sigra ward are on private land. Few dumps in Nagwa ward, Jaitpura ward, Sicrol ward and Sarnath ward are also on private land. On the bank of the River Ganga, there are 9 sites where waste is being dumped. All the religious activities which are being carried out along the Ganga River result in waste generation, which is dumped along the river; a large proportion of which is flowers and waste generated from the temples. On the other hand, wastes through the hospitals and nursing homes are highly toxic as it contains hazardous materials endangering our natural environment. As for as the proper collection and disposal of hospital wastes are concerned the Nagar Nigam has given clear cut instructions to the hospitals, nursing homes and pathological centres of the city to manage their own system for scientific disposal of the wastes. Sorry to say that have of them complied over the suggestion of the Nagar Nigam.

References

Ahmed, S., Sayed, M. H., Barun, S., Khan, M. H., Faruquee, M. H., & Jalil, A. (2001). Arsenic in drinking water and pregnancy outcome. *Environmental Health Perspectives, 109*, 629–631.

Krishnan, M. S. (1960). *The geology of India and Burma*. Mount Road, Madras: The Associated Printers (Madras) Private Ltd., p. 555.

Mc Bean, E. A., Rovers, F. A., & Farquhar, G. J. (1995). *Solid waste landfill engineering and design* (p. 380). Englewood Cliffs: Prentice Hall.

Mohan, K., Shrivastava, A., & Rai, P. K. (2011). Ground water in the city of Varanasi, India: Present status and prospects. *Question Geographicae, 30*(3), 47–60. doi:10.2478/v10117-011-0026-9.

Pathak, B. D. (1977). *Geology and ground water conditions of Varanasi district*, U.P. Ball. Geological Survey of India Series (B), No. 41.

Rai, P. K., & Mohan, K. (2014). Remote sensing data & GIS for flood risk zonation mapping in Varanasi district. *Forum Geographic Journal (Romania)*, 13(1), 25–33. Available at doi:http://dx.doi.org/10.5775/fg.2067-4635.2014.041.i

Rana, P., Singh, S. K., & Banerjee, H. (2003). Organochlorine pesticide residue in groundwater in world oldest existing civilized city, Varanasi. *Indian Journal of Agricultural Environment & Biotechnology, 1*, 94–107.

Report of the 'Hindustan' Newspaper, a Hindi dailies, Ground water at an alarming stage. Feb. 27, 2006, 3.

Roychowdhury, T., Basu, G. K., Mandal, B. K., Biswas, B. K., Chowdhury, U. K., & Chand, C. K. (1999). Arsenic poisoning in the Ganges Delta. *Nature, 401*, 545–546.

Shah, N. S. (2007). Burning quest: Should public health be sacrificed to waste? *Down To Earth,* 25–26.

Sharholy, M., Ahmad, K., Vaishya, R., & Gupta, R. (2007). Municipal solid waste characteristics and management in Allahabad, India. *Waste Management, 27*(4), 490–496.

Singh, S. K., Raha, P., & Banerjee, H. (2006). Banned organ chlorine cyclodeine pesticide in groundwater in Varanasi, India. *Bulletin of Environmental Contamination and Toxicology, 76*, 935–941.

Suchitra, M. (2007). *Outside: Burnt or buried, garbage needs land.* Down To Earth, pp. 22–24.

Chapter 3
Integration of Census Data with GIS for Analysis of Population Characteristics

Abstract GIS is very well used to integrate the census data for population characteristics. Analysis of population characteristics plays an important role in population geography. In this chapter various aspect of population characteristics like population growth, distribution, density, sex ratio, literacy, occupational structure of peoples in Varanasi are discussed. Population characteristics are relatively vital because it defines the economic and cultural condition as well as characteristics of the area, concerned. The decennial growth rate of Varanasi district had been very high during the last three decades. This has resulted in the pressure of population on land. Growth rate of male-female population during last 3–4 decades (from 1971 to 2001) in Varanasi district has been increased more than double. As per the census 2001, the current sex ratio (female population per 1000 male) in Varanasi city is 876, which is lower than the state urban average of 885 and national urban average of 901. Such rural-urban differentials in sex-ratio are the product of sex selective migration from rural areas to urban areas. Occupational composition is an important index to through light upon the economic dynamic of health and vigor of a region. The growth in population is also likely to pressure already stressed public transport and will have effect on health services, hence planned efforts are required to direct the growth of the district both rural and urban area in right direction.

Keywords GIS • Population characteristics • Population growth • Population density • Sex ratio • Literacy

3.1 Population Characteristics

Analysis of population characteristics plays a very significant role in population geography. Population characteristics are quite important because it defines the economic and cultural condition as well as characteristics of the area, concerned. Age and sex structure are the two important physiological characteristics which play an important role in any study of population problem as they determine future potentials of growth of population to a very large extent (Verma 1967). They affect not only the demographic characteristics but also the socio-economic and political structure of any region (Thomlinson 1965). Population resource is a factor of prime

P.K. Rai, M.S. Nathawat, *Geoinformatics in Health Facility Analysis*,
DOI 10.1007/978-3-319-44624-0_3

importance in the economic development of any country. The man, infact is the resource of all the resources and also the beneficiary in the complete process of the resources development and utilization. Hence, there is a great need for a systematic study of population in many aspects, because human resource gives a definite line of action to exploit with population characteristics and processes both the drivers and results of social and economic development processes and outcomes. It is authoritative that a good understanding of a country's population dynamics provides the basis for informed decision-making, policy development and planning (Kinsley 1959).

Man is the producer, investor and consumer of resources and thus, he is not only the beneficiary of the complete process of resource development and utilization, but is also the most effective and dynamic agent of production (Ezimmermann 1951). Hence, population is the essential element from which resource features originate their significance and as such human resources, may play an important role in its economic growth. Therefore, the study of population as a resource requires special consideration in any development planning. The study-area, in spite of its spatio-temporal variation in physiography, rainfall and other physical factors, is characterised with the variety of food and commercial crops with introduction and expansion of irrigation facilities adoption of agricultural revolutions. Obviously majority of the inhabitants of the district depend for their livelihood on their valuable land resources. Due to fast increase in population during earlier decades population pressure on land has become very high which is adversely affecting the economic development of the area. Therefore, in the part of the chapter emphasises is laid on the assessment of human resources with references to its spatial distribution, growth, density, literacy, sex ratio (Chandrashekhar 1967). The pressure of population on land and other resources or facilities like health facilities becomes noticeable when we study occupational structure. Its study is of vast significance for planning and development of human resource. This important attribute of population exert vital effect on several socio-economic personalities of the workers. Perhaps, none of the facets of a population throws as much light on the economic growth of region as its occupation structure (Sinha and Zacharia 1979).

3.2 Population Projection

Planned economic development requires data about various aspects of socio-economic conditions at different levels. Indicators of development are directly or indirectly related to the size and structure of the population. It is, therefore, of paramount importance to know various aspects of the size and structure of population at different points of time (Chandana and Sidhu 1980; Chandna 1986).

Population projection is important and basic requirement for the provision of basic services to the people. It is also required to plan for service provision and facilities analysis for the people in the district, which is the direct function of the population and population growth. A projection may be made by a governmental organization, or by those unaffiliated with a government. They refer mostly to the exercises of extrapolation of the past trends into the future; and they do not take into account changes in the policy parameters. For example, a projection of the future population growth may not be taking into account changes in the government health policies, family health care programmes, etc. Projections are based on the assumption that the past trends will continue to operate in the future. The reliability and usefulness of projections depend on the assumptions and their closeness to reality (Demco et al. 1970). In the long run, the policy parameters are to be incorporated in the projections. The likely effects of policy changes are to be arbitrated and projections are to be prepared accordingly. Thus, when an element of judgment is added to the projections, it becomes a forecast. Forecasts enjoy the advantage of being based upon the assumption or a set of assumptions which are likely to be realized in the near future and can yield a relatively more realistic picture of the future.

The Varanasi district, both its urban and rural areas has an exceptionally different growth character, accomplished by the movement of people from surrounding areas for occupational reasons, tourist traffic as a result of its heritage value, and special events of mystical importance of the Ganga river at Varanasi. The proximal townships of Ramnagar and Mughalsarai, both of which lie across the Ganga river, east of Varanasi, accompaniment the growth of the Varanasi region. The base data used for population projection is the data acquired from the Census of India, with detailed urban area population and district population for 2001 and the 1991 census data summaries (Table 3.3). This data provided the numeric basis for benchmarking the actual population and its decadal growth for the past decades. There are various methods of projecting population (mathematical, economic and component methods). Some are very sophisticated and rigorous while others are simple and less sophisticated. Normally, population in future is governed by the following equation:

$$\mathrm{P_n = P_o + Number\ of\ Births(B) - Number\ of\ Deaths(D) + Net\ Migration(N_m)}$$

For the projection of population in 2010 (P_n), base year population (P_o) in 2001, the number of births and deaths between 2001 and 2010 and net migration is required. Keeping in view of the in-migration and out-migration, net migration may be either positive or negative.

One such method that considers all these aspects is known as component method of population projections. This process is suitable for projecting sub-national i.e. district and block level population. It requires detailed age-structure of population in the base year along with estimation of variability of demographic indicators.

For computing the annual rate of growth (simple), the following formula can be applied to the information at any two points of time.

$$r = \frac{1}{n}\frac{[Pn - Po]}{Po}x100$$

Where, r = annual rate of growth
P_n = population in the current year
P_o = population in the base year
n = number of intermediary years.

A slightly improved method is the compound rate of growth method, which can be computed with the help of the following formula.

$$R = \left[\left(P_n / P_o \right)^{1/n} - 1 \right] \times 100$$

By the formula

$$P_n = P_o \left(1 + R / 100 \right)^n$$

3.3 Population Density

The overall population distribution in the district is closely related to the physical and socio-cultural features. Population distribution is a dynamic process which displays varying nature of man's adjustment with physical resources.

3.3.1 Arithmetic Density

The most common type of population density is the arithmetic density expressed in terms of ratio of the total population to the total area. For the present analysis block-wise arithmetic density has been worked out which reveals an irregular and uneven distribution of population density of 997 persons per sq. km. while it varies from 1266 persons per sq. km. in Kashi Vidyapith to lowest of 877 persons per km sq. Km in Pindra Development Block. Table 3.1 and Fig. 3.1a, b show that the arithmetic density in the study area of year 1991 and 2001, which has been largely influenced by urban centres and terrain conditions.

3.3.2 Physiological Density

The ratio between total population and cultivated land of a region is termed as the physiological density. This signifies the pressure of population on agricultural land which is the main source of livelihood for rural masses. The physiological density of 1287 persons per sq. Km. for the district as a whole exhibits marked spatial variation form more than 1400 persons per sq. Km. in Kashi Vidyapith and Harhua

Table 3.1 Population density in Varanasi district, 2001

Development block	Arithmetic density (sq. km.)			Physiographic density (sq.km.)			Agriculture density (sq.km.)			Rural density (sq.km.)
	1981	1991	2001	1981	1991	2001	1981	1991	2001	2001
Baragaon	774	966	1124	996	1235	1497	143	233	206	1124
Pindra	687	903	1068	815	1092	1329	134	246	165	1067
Cholapur	655	840	1139	881	1114	1404	134	238	154	1139
Chirai Gaon	743	1037	1270	957	1258	1683	164	209	131	1270
Harhua	878	1173	1531	1157	1499	2034	164	302	192	1532
Sewapuri	739	964	1241	950	1262	1541	150	231	194	1211
Araziline	842	1110	1424	1043	1363	1739	196	226	190	1425
Kashi Vidhyapith	859	1186	1797	1237	1614	2451	199	219	131	1797
Total Block	10350	1014	2044	1005	1287	1326	160	237	280	1426

Source: Census of India, Primary Census Abstract (PCA), Varanasi district, 1981, 1991 and 2001 and self computed

blocks to less than 1100 persons per sq. km. in Pindra development block (Table 3.1 and Fig. 3.2b). Normally, the high physiological density is imitated by high intensity of cultivation.

3.3.3 *Agricultural Density*

Agricultural density indicates the relationship between persons involved and the agricultural area of any region. Crude density is not the measure of crowding in agricultural countries like India. It is the agricultural density which might be more suitable to signify the extract number of persons directly dependent on agricultural per sq. km. of cultivated area.

The district as a whole, observed an agricultural density of 237 persons per sq. km. (Table 3.1). With reference to agricultural density the Varanasi district can be grouped into five categories i.e. very low, low, medium, high and very high (Fig. 3.2c). Very high density (above 260 persons per sq. km.) has been detected in Harhua development block while very low density (below 215 persons per sq.km.) marked in Chiraigaon development Block. Pindra development Block shows high agricultural density while Barragaon, Sewapuri and Cholapur development Blocks come under medium and Kashi Vidyapith and Araziline Blocks witness low agricultural density.

Agricultural density is mostly affected by the presence of household industries and urban centres. It denotes per farm worker's availability of the arable lands, the high agricultural density may be observed in those development blocks where number of farm workers is low and share of arable land is high (e.g. Harahua and Pindra blocks). On the otherhand, low density is identified in those blocks where the

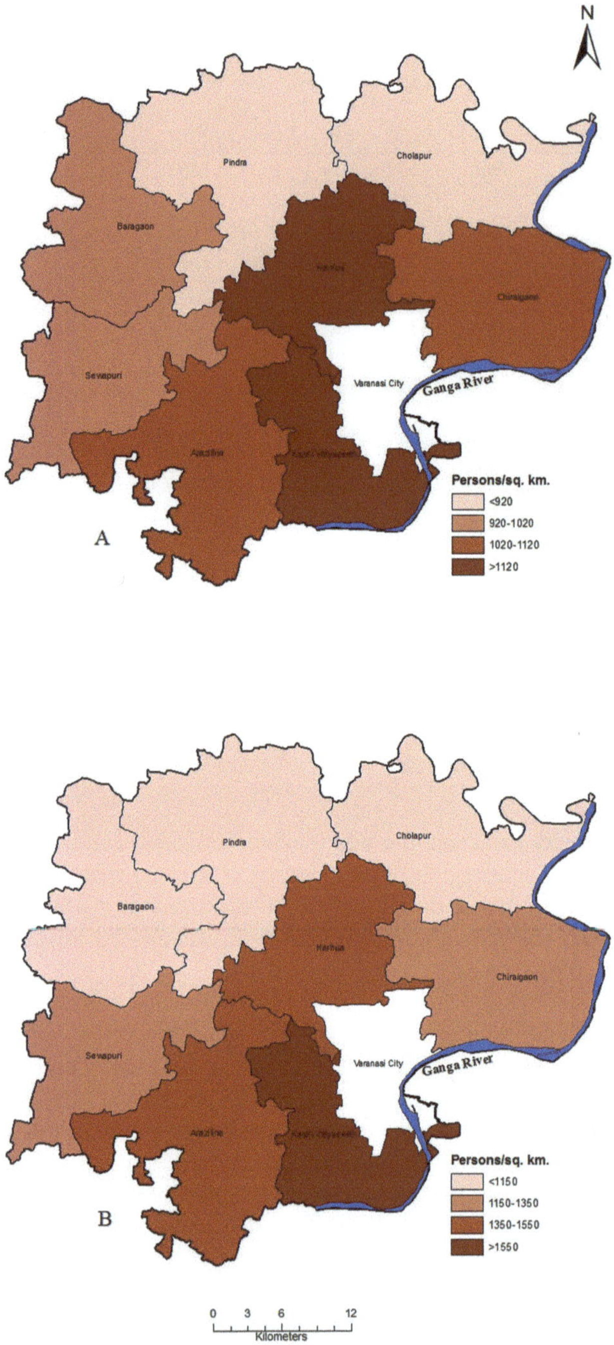

Fig. 3.1 Arithmetical density in the year 1991 (**a**) and 2001 (**b**) in rural area of eight development blocks of Varanasi district respectively

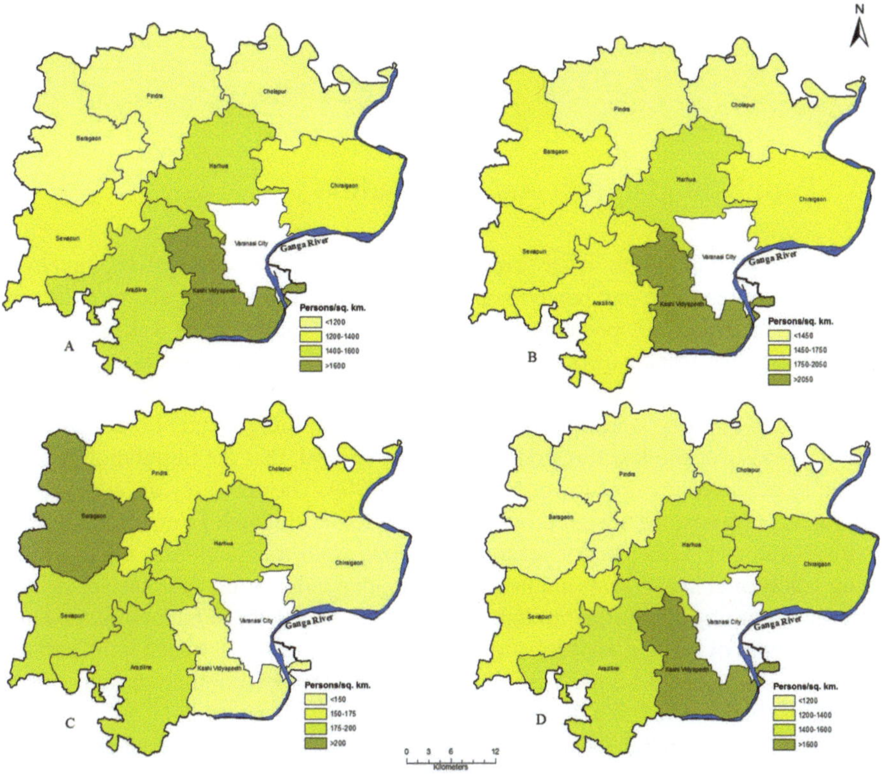

Fig. 3.2 Rural (**a**), physiological (**b**), agricultural (**c**) and economic density (**d**) in eight development blocks of Varanasi district respectively, 2001

Table 3.2 Religion-wise populations in Varanasi district, 2001

Religious community	Population			% in total population
	Total	Rural	Urban	
Hindu	26,27,565	17,62,385	86,5180	83.72
Muslim	49,7516	11,1381	38,6135	15.85
Christian	44,99	10,76	3423	0.14
Sikh	44,96	13,81	31,15	0.14
Buddhist	985	476	509	0.03
Jain	17,47	291	14,56	0.06
Other	167	20	147	0.01
Unspecified	16,96	1090	606	0.06
Total	3138671	1878100	1260571	100.00

Source: Census of India, Primary Census Abstract (PCA), Varanasi district, 2001 and self computed number of farm workers is more and arable land is less (e.g. Kashi Vidyapith and Araziline blocks). High agricultural density may be characterised with less urbanization and low pressure on arable land.

3.4 Population Growth and Distribution

The growth of population in any area whether it is positive or negative, generally shows human response and behavior to his environment suitability and possibilities (Thomlinson 1965). To speculate the future with some confidence it is essential to evaluate the past trend of population growth. Looking at the decadal growth rate variations from 1981 to till 2001, one can conclude that there are fluctuations in decadal growth rates, and since 2001 it reflects declining trend. Looking at the growth trend over the last three decades, it can be said, that the population projections thus achieved are on a lower side in 2008–2009. The declining decadal growth rate for Varanasi can be attributed primarily due to the sluggish growth of industrial sector of the city. Other major reasons contributing to this declining trend are the dying traditional small-scale industries, heavy dependence on tourism sector and the crumbling physical infrastructure of the city. Keeping these factors in view and with the intent of being more realistic in population projections, which will form the basis of subsequent design efforts; these higher endpoint populations have been used as the basis of refining the population growth and dispersion.

Table 3.2 and Fig. 3.3 show the religion-wise population in Varanasi district. As per census 2001, it was found that 83.72 % of the total population is belonging to Hindu community whereas 15.36 % belong to Muslim. The decennial growth rate of Varanasi district had been very high during the last three decades. This has resulted in the pressure of population on land. During the last two decades (1971–1991) the

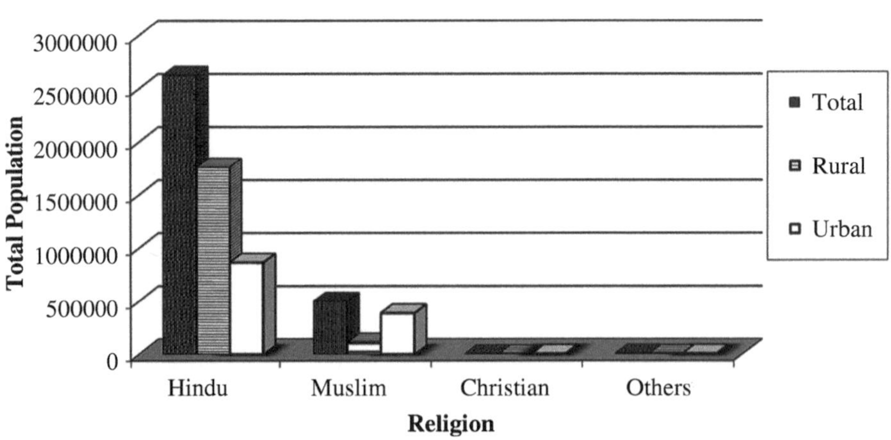

Fig. 3.3 Religion-wise population in Varanasi district, 2001

Table 3.3 Population growth rate (%) in Varanasi district

Development block	Population 1981	1991	2001	2010[a]	Growth rate (%) (1981–1991)	Growth rate (%) (1991–2001)	Projected growth rate (%) (2001–2010)
Baragaon	129744	161843	195972	234053	24.74	21.08	19.43
Pindra	149215	196025	238627	262878	31.37	21.73	11.78
Cholapur	126085	162185	204348	234566	28.63	25.99	14.78
Chirai Gaon	139932	185521	248894	267878	32.58	34.15	7.62
Harhua	124596	166466	216189	240658	33.6	29.86	11.31
Sewapuri	121638	158541	204731	229340	30.34	29.13	12.02
Araziline	177973	234616	311723	340299	31.83	32.86	9.16
Kashi Vidhyapith	128683	184941	257616	267050	38.07	39.29	3.66
Total Block	1097866	1450138	1878100	2056722	31.66	29.51	9.51
Urban	773865	1030863	1313242	1500801	33.21	16.64	14.28
Total district	1871731	2481001	3147927	3557523	32.55	26.88	13.01

Source: Census of India, Primary Census Abstract (PCA), Varanasi district, 1991 and 2001 and self computed
[a]projected population, 2010

total population of the region has increased by 65.27 % while decennial growth rate (1981–1991) for the district as a whole was 32.55 % whereas 26.88 % in year 1991–2001. Kashi Vidyapith (38.07 % in 1981–1991 and 39.29 % in 1991–2001) and Chiraigaon (32.58 % in 1981–1991 and 34.15 % in 1991–2001) blocks recorded higher growth rate than the district average whereas Baragaon (24.74 % in 1981–1991 and 21.08 % in 1991–2001) block completely rural in character witnessed lower growth rate (Table 3.3 and Fig. 3.4). It might be attributed to rural to urban migration.

Based on the assumption that the same declining trend has also continued till 2010–2011, present population should be around 35, 57,523 and considering the decadal rate at 13.01 %. Baragaon development block shows higher growth rate during 2001–2010 time periods i.e. 19.43 % followed by Cholapur and Sewapuri development respectively blocks, which are 14.78 % and 12.08 % (Table 3.3).

Again based on this population projection assumption for Varanasi city area, the same declining trend has and will continue till 2011, 2008 population should be around 1,500,801, and considering the decadal rate at around 14 %, population for 2011 should be around 15, 35,279, population rate should increase over next decades and reach 28 % by 2021 and 31 % by 2031. A small increase of 3 % is assumed for year 2031 keeping in mind that the satellite towns like Ramnagar and Mugal Sarai will also be developed and there will be less developmental pressure on Varanasi. Reanalyzing the data for population projections, it can be estimated that population for year 2021 should be around 19, 65,157 and for year 2031 it should be around 25, 72,356 (Table 3.4). The growth in population is also likely to stress already stressed

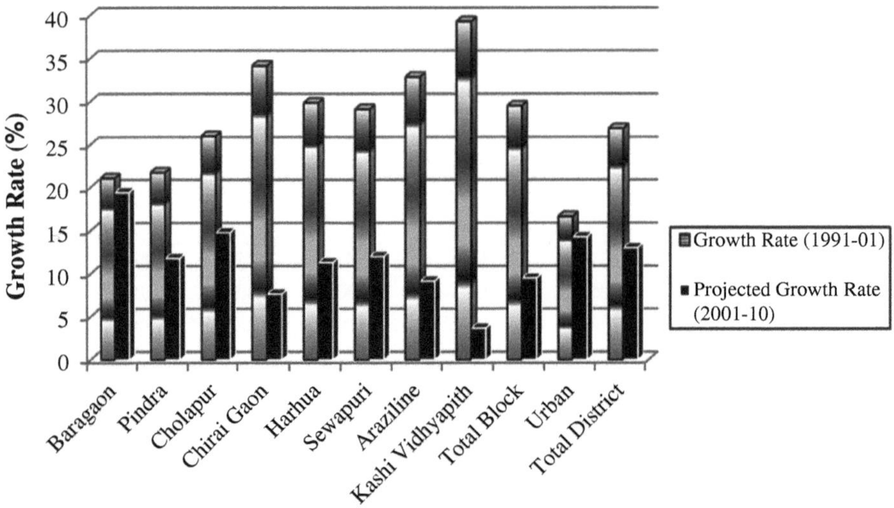

Fig. 3.4 Population growth rate in Varanasi district

Table 3.4 Population projections on decadal growth rate of Varanasi city

Year	Population	Decadal growth rate (%)
1961	4,89,864	37.69
1971	6,71,934	26.14
1981	7,73,865	25.23
1991	10,30,863	33.21
2001	12,02,443	16.64
2006	13,70,785	14.00
2011	15,35,279	12.00
2021	19,65,157	28.00
2031	25,74,356	31.00

Source: Census of India, Primary Census Abstract (PCA), Varanasi district, 1991 and 2001 and self computed

public transport and will have impact on health services, hence planned efforts are required to direct the growth of the district both rural and urban area in right direction. Population distribution of Varanasi district for eight development block of year 2001 and 2010 are shown in the Figs. 3.5 and 3.6 respectively. In this each dot represents 500 persons living in the Varanasi district.

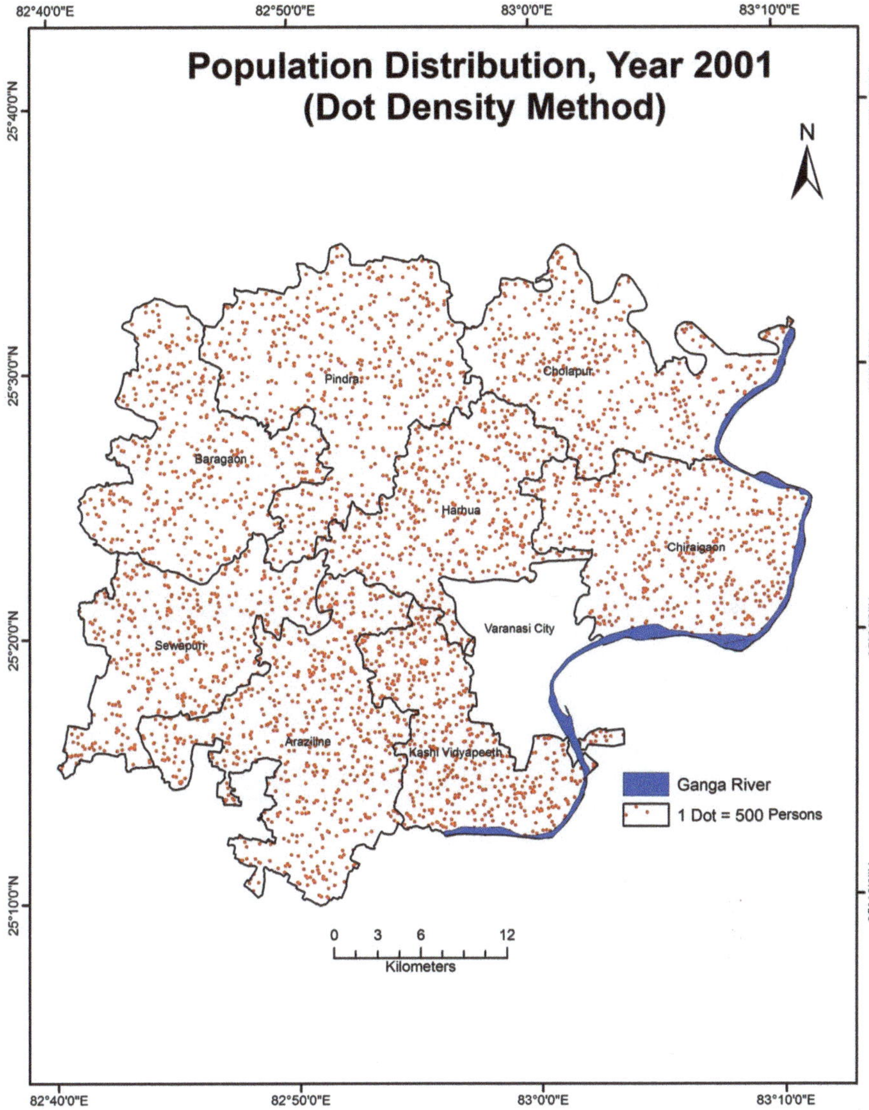

Fig. 3.5 Population distribution of eight rural development block of Varanasi district, 2001

3.4.1 Male-female Population Growth Rate

Growth rate of male-female population during last 3–4 decades (from 1971 to 2001) in Varanasi district has been increased more than double. Higher male population growth-trend is exemplified by Baragaon (80.8 %) development block followed by Pindra (80.4 %) and Harahua (77.5 %) respectively. Low male population growth is perceived in Chiraigaon (74.6 %) development block. In comparison to male

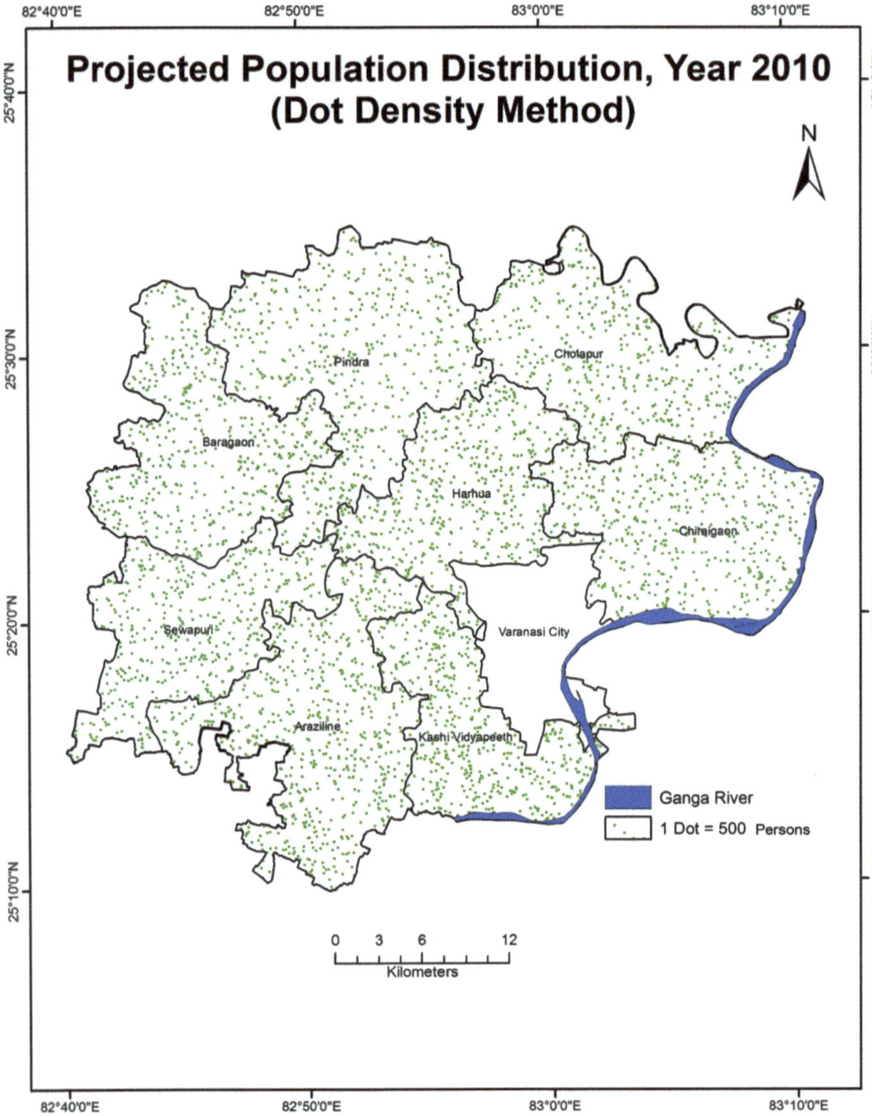

Fig. 3.6 Projected population distribution of eight rural development block of Varanasi district, 2010

Table 3.5 Male and female population growth rate (in %)

Development block	Growth rate in % (Male population)				Growth rate in % (Female population)			
	1971	1981	1991	2001	1971	1981	1991	2001
Baragaon	40.0	48.9	69.0	80.8	8.0	12.0	26.3	47.0
Pindra	39.0	47.1	68.0	80.4	9.0	11.3	25.4	45.9
Cholapur	35.0	43.9	61.6	75.3	8.0	4.9	23.2	45.1
Chirai Gaon	32.0	38.5	54.7	74.6	8.4	9.1	18.1	44.6
Harhua	34.0	43.5	62.5	77.5	8.2	9.6	22.6	46.5
Sewapuri	37.0	43.8	64.4	77.3	5.0	11.7	23.9	44.3
Araziline	36.0	42.8	62.3	75.2	9.0	8.4	21.4	43.1
KashiVidhyapith	31.0	41.8	60.6	76.4	8.6	12.2	26.9	47.0

Source: Census of India, Primary Census Abstract (PCA), Varanasi district, 1991 and 2001 and self computed

population, very low growth was recorded in female population during the census period of 1971–1981, 1981–1991 and 1991–2001. During 1991–2001, the population growth rate of female in Baragaon and Kashi Vidyapith development blocks increased by 47 % and in the decades, 1971–1981 and 1981–1991, this increase were 12 % and 26.3 % for Baragaon and 12.2 % and 26.9 % for Kashi Vidyapith development blocks respectively (Table 3.5, Figs. 3.7 and 3.8).

3.4.2 Scheduled Caste/Scheduled Tribe (SC/ST) Population

The Scheduled Caste population contribution very much to the traditional rural agricultural economy, provide the major working force as hired agricultural laborer. SC/ST population generally forms a major share of backward society (Ghosh 1985; Ghosh and Mukharjee 1989). They live in different socio-economic environment and hence, present some typical demographic characteristics. Even their different groups (social and economic) represent different antecedents and level of techno-economic development. Of the total population, nearly 13.84 % is that of schedule caste and schedule tribes in 2001 while 13.98 % in 1991. Amongst the development blocks there is found much spatial variation in distribution of SC/ST population. Cholapur development block recorded the highest population (22.94 %) of SC and ST followed by Chiraigaon (20.83), Pindra (18.89 %) and Baragaon (18.46 %). The medium proportion (15–18 %) is observed in Kashi Vidyapith, Harhua and Sewapuri blocks. The Araziline blocks falls in the lowest category of SC/ST population of the total population (13.72 %) in 2001 but this figure was 13.8 % in 1991, which is approximately 7.4 % of the total Municipal area population. As per the census 2001, ST population in the city stands at 483, which is a simple 0.04 % of the municipal population (Table 3.6 and Fig. 3.9). The sex ratio for the ST population in the city stands at 1021, which is much higher than the total sex ratio of the city. The SC population in Varanasi Municipal area is 81,704 (Census of India 2001).

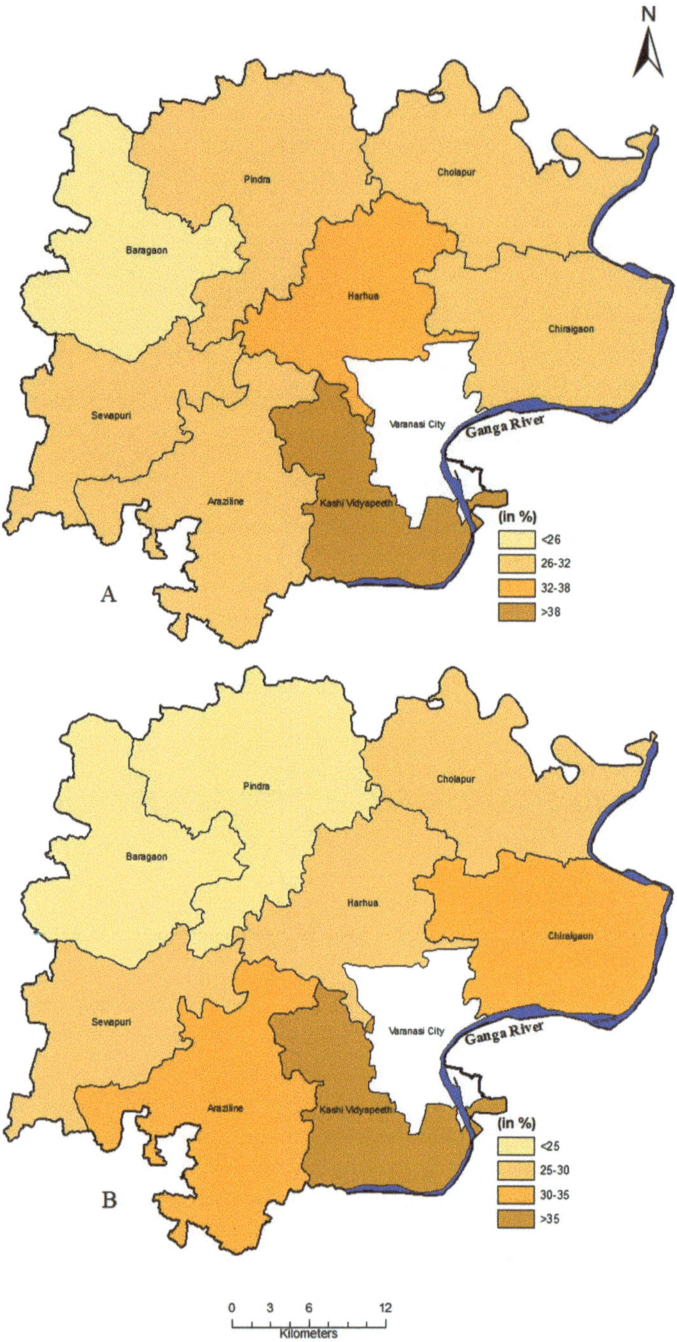

Fig. 3.7 Female population growth rate in rural area of each development blocks of Varanasi district in year 1981–1991 (**a**) and 1991–2001 (**b**) respectively

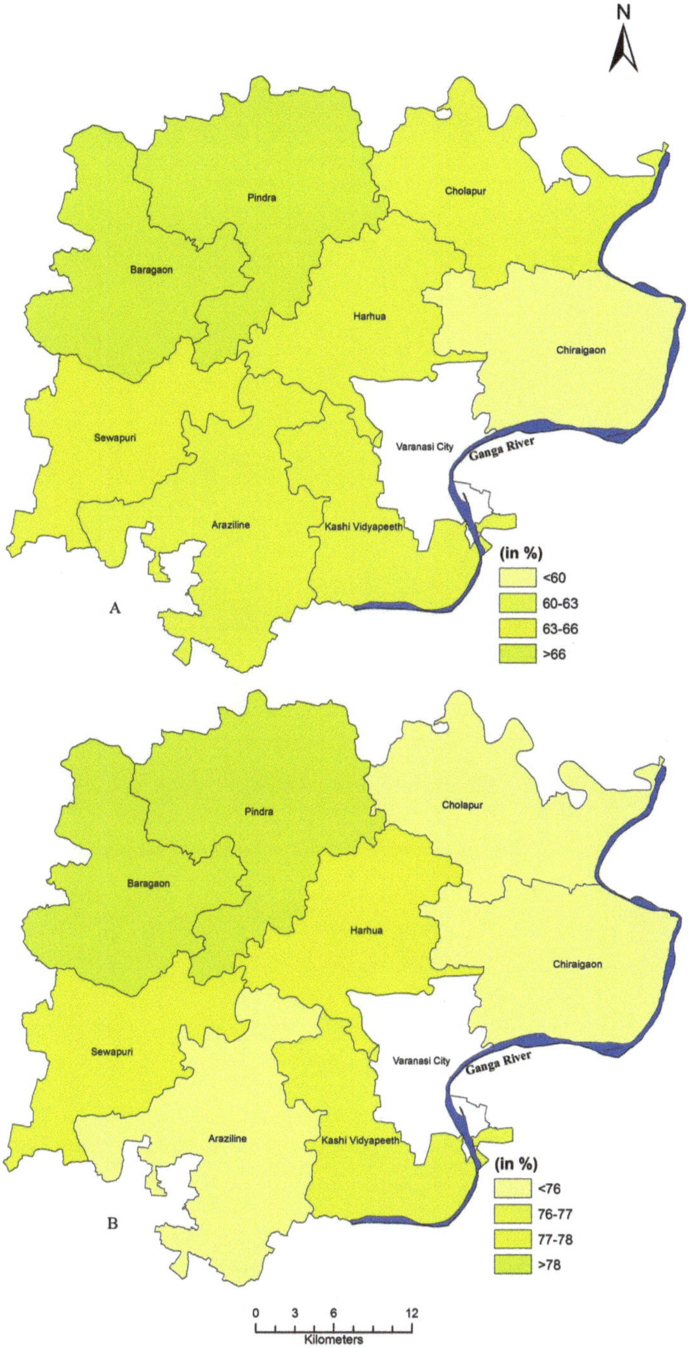

Fig. 3.8 Male population growth rate in rural area of each development blocks of Varanasi district in year 1981–1991 (**a**) and 1991–2001 (**b**) respectively

Table 3.6 Percentage of schedule caste/schedule tribe (SC/ST) population to the total population

Development block	SC/ST population, 1991			% of SC/ST population to total population, 1991	SC/ST population, 2001			% of SC/ST population to total population, 2001	Change: 1991–2001		
	Male	Female	Total		Male	Female	Total		Male	Female	Total
Baragaon	14861	14519	29380	18.15	18108	18069	36177	18.46	21.85	24.45	23.13
Pindra	18421	17524	35945	18.34	22929	22158	45087	18.89	24.47	26.44	25.43
Cholapur	19903	17809	37712	23.25	24517	22440	46957	22.94	23.18	26.00	24.51
Chirai Gaon	20119	17925	38044	20.51	27382	24531	51913	20.83	36.10	36.85	36.46
Harhua	15443	13936	29379	17.65	20234	18123	38357	17.73	31.02	30.04	30.56
Sewapuri	14459	13214	27673	17.45	18995	17456	36451	17.80	31.37	32.10	31.72
Araziline	17262	15221	32484	13.85	22628	20162	42790	13.72	31.09	32.46	31.73
KashiVidhyapith	14715	12554	27269	14.74	19471	17010	36481	14.16	32.32	35.49	33.78
Total Rural	135183	122702	257885	17.78	174264	159949	334213	17.78	28.91	30.36	29.60
Total Urban	48033	41046	89079	8.64	54892	47209	102101	7.73	14.28	15.01	14.62
Total District	183176	163718	346894	13.98	229156	207158	436314	13.84	25.10	26.53	25.78

Source: Census of India, Primary Census Abstract (PCA), Varanasi district, 1991 and 2001 and self computed

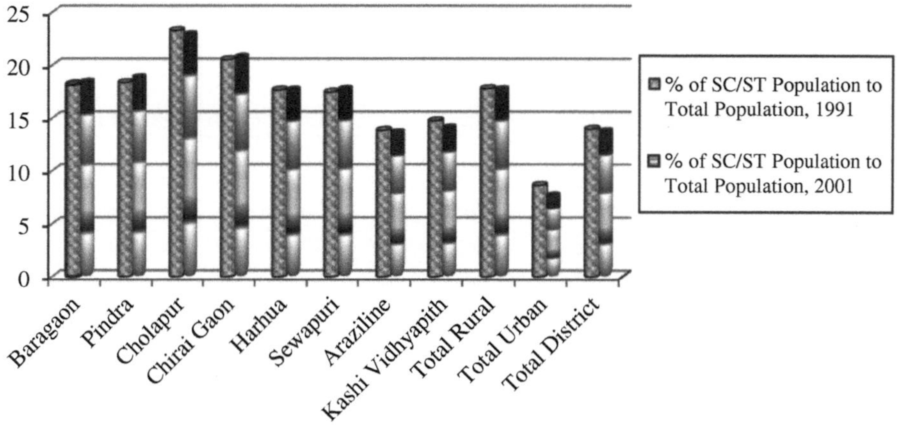

Fig. 3.9 SC/ST population to total population of Varanasi district

3.5 Spatial Pattern of Sex Ratio

Sex ratio is an index of the socio-economic conditions leading in an area and is valuable tool for regional study. It has a profound consequence on other demographic elements (Franklin 1956). The sex ratio is generally specified as the number of females per thousand males. It is an indicator of social and economic development and development of an area and a significant tool for regional study (Trewartha 1953; Chandana and Sidhu 1980; Chandna 1986). There are three type of sex ratio: primary sex ratio (computed at the time of conception), secondary sex ratio (computed at the time of birth), and tertiary sex ratio (computed at the time of enumeration). The sex ratio of the district was 890 in 1991 which increased to 908 in 2001. Block-wise distribution varies from 960 females per thousand males the highest in Baragaon to the lowest of 868 in Kashi Vidyapith which lies in the environs of the Varanasi city (Table 3.7 and Fig. 3.10). It clearly reveals from the Table 3.7 that 29.51 % of the decadal increment in rural population of Varanasi district form year 1991–2001.

Table 3.7 Sex ratio of total population in Varanasi district

Development Block	Rural male population (2001)	Rural female population (2001)	Total rural population (2001)	Decadal increment 1991–2001 (%)	Total sex ratio				Change in sex ratio: 1991–2001 (%)
					1971	1981	1991	2001	
Baragaon	98758	97214	195972	21.09	1010	983	856	983	2.82
Pindra	12723	117904	238627	21.73	978	980	957	976	1.99
Cholapur	105189	99159	204348	26.0	940	948	918	942	2.61
Chirai Gaon	130916	117978	248894	34.16	899	912	894	901	0.78
Harhua	113223	102966	216189	29.87	892	903	908	909	0.11
Sewapuri	106098	98633	204731	29.13	975	948	919	929	1.09
Araziline	164119	147604	311723	32.87	910	900	887	899	1.35
Kashi Vidhyapith	137029	120587	257616	39.3	867	856	868	880	1.38
Total District	976055	902045	1878100	29.51	866	872	890	908	2.02
Total Urban	–	–	–	–	844	854	862	876	1.62

Source: Census of India, Primary Census Abstract (PCA), Varanasi district, 1991 and 2001 and self computed

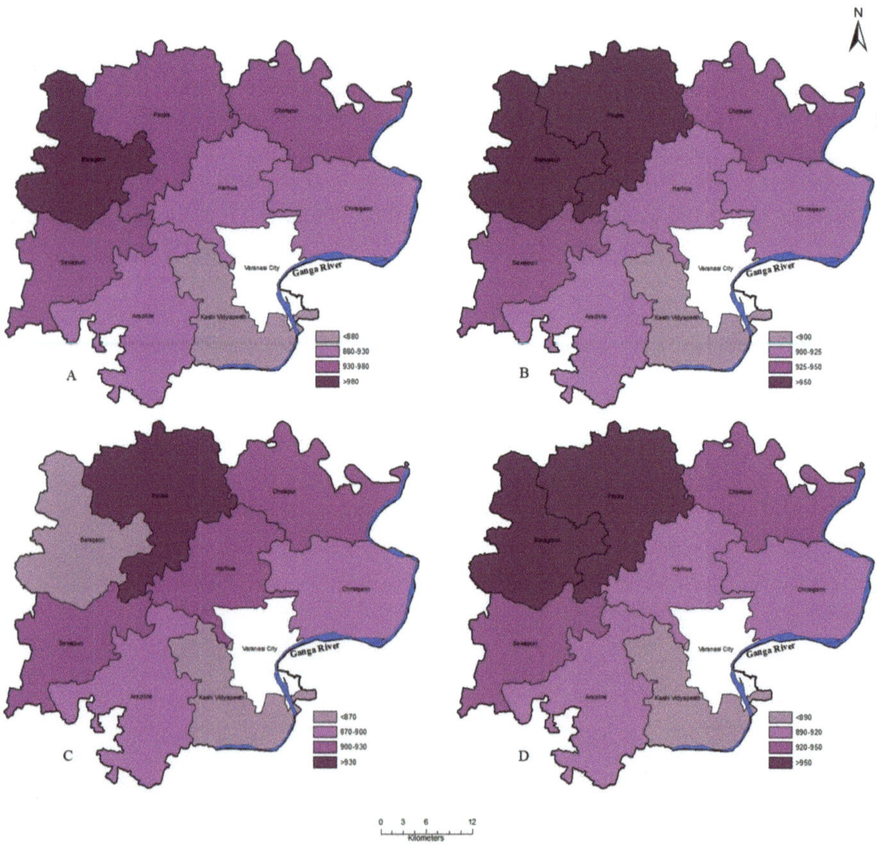

Fig. 3.10 Sex ratio in Varanasi district of 1971 (**a**), 1981 (**b**), 1991 (**c**) and 2001 (**d**) respectively

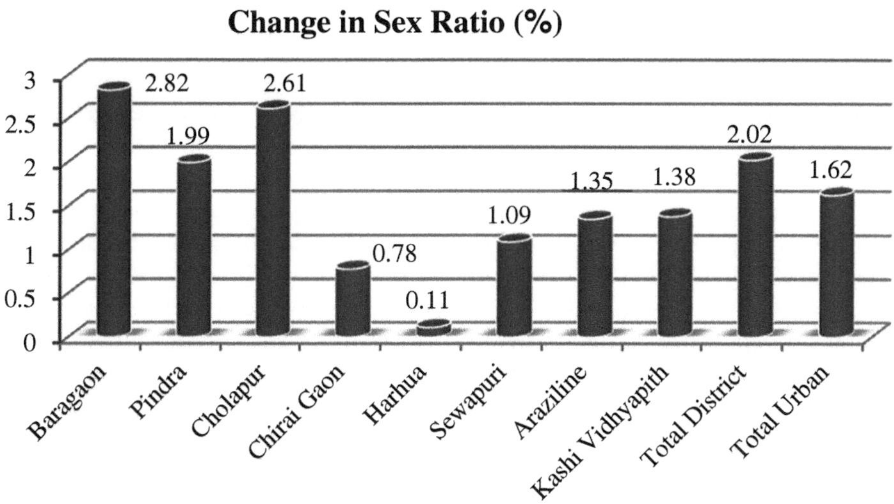

Fig. 3.11 Change in sex ratio (%) in Varanasi district

Altogether, there are three development blocks which belong to the high sex-ratio group of above 950 females per 100 males. Two blocks, namely Harhua and Cholapur fall in the medium group 900–950 whereas three blocks come in the group of sex-ratio below 900. Low sex-ratio is observed in Chiraigaon, Kashi Vidyapith and Araziline blocks due to their proximity to Varanasi city where a large number of people live in rented houses located in the block to perform behind their duties leaving their family members at their native place. Varanasi district is characterized by wide difference in its rural-urban components. The rural areas of the district have higher sex ratio in comparison to the urban areas. As per the census 2001, the current sex ratio (female population per 1000 male) in Varanasi city is 876, which is lower than the state urban average of 885 and national urban average of 901. Such rural-urban differentials in sex-ratio are the product of sex selective migration from rural areas to urban areas. The increasing pressure of population upon the limited agricultural resource base in the countryside compels the rural males to move to urban areas in search of jobs. It is remarkable to note that from Table 3.7 and Fig. 3.11, according to 1991 and 2001 census, Varanasi district shows 2.02 % change in sex ratio during this period, in which Baragaon development block shows highest (2.82 %) changes in sex ratio (856 in 1991 and 983 in 2001).

3.6 Spatial Pattern of Literacy

Literacy in the best indicator for economic development, social advancement and demographic growth of the region. Literacy is both the cause and the effect of development (Chandana and Sidhu 1980; Chandna 1986). Education is double-edged instruments which can eradicate the effects of socio-economic inequalities, but which can itself introduce a new kind of inequality between those who have it and those who do not. Thus, the period of economic development and literacy pattern of a region or area are closely related (Clark 1977). There have been numerous studies in the past on literacy in the Indian context. These have dealt with disparities between Scheduled caste and non-Scheduled caste population pointing out that disparities increase from urban to rural areas and are more pronounced among specific females groups in rural areas and the deprivation of women in the field of education (Chandana and Sidhu 1980; Chandna 1986).

Census of India treats person aged seven and above, who can both read and write with understanding in any languages, as literate. A person, who can only read but cannot write, is not literate (Census of India 2001). Prior to 1991, children below 5 years of age were essentially treated as illiterates.

The ability to read and write with understanding is not ordinarily achieved until one had some schooling or at least some time to develop these skills. It was, therefore, decided at the 1991 Census that all children in the age group 0–6 years, will be treated as illiterate by definition and the population aged 7 years and above only be classified as literate or illiterate. The same criterion has been retained at the Census of India 2001, also. It is not mandatory that to be treated as literate, a persons should have received any formal education or acquired any minimum educational standard. Literacy status can be acquired through literacy classes or by attending any non-formal educational system. Persons who are unfortunately blind and read in Braile are also treated as literates. Till 1981, it was customary to work out the literacy rate taking into account the total population. Since literacy rate is more meaningful if the sub-population in the age group 0–6 years is excluded for the total population, it was decided in 1991 to use the term literacy rate for the population relating to 7 years and above. The same concept has been continued for Varanasi district. The literacy rate taking into account the total population in the denominator has now been termed as 'crude literacy rate' which is as follow:

$$\text{Crude Literacy Rate}\left(\text{CLR}\right) = \frac{L7+}{P} * 100$$

$$\text{Literacy Rate}\left(\text{LR}\right) = \frac{\text{L7+}}{\text{P7+}} * 100$$

Where,
L7+ = Number of Literate population aged 7 and above
P7+ = Number of total population aged 7 and above
P = Number of total population.

A number of socio-economic and traditional factors are accountable for variation in literacy rates in different parts of the region. Social responsiveness and other institutional infrastructure are other factors to promote literacy in the Varanasi district. In Varanasi district proportion of literacy persons expressed in percent to the total population is 43.8 % in 1991. Across districts, Varanasi district has moderate literacy rates (67.12 %, as per 2001 census). Current total literacy rate within Varanasi city is high (77.87 %) as compared to state urban average of 56.3 % and national urban average of 70.1 %.

The highest literacy rate (above 63 %) has been observed in Baragaon, Pindra and Kashi Vidyapith developments blocks while comparatively medium literacy (61–62%) is recorded in Harahua and Sewapuri blocks. The lowest literacy rate i.e. around 60 % is recorded in Cholapur, Chiraigaon and Araziline development block (Table 3.8 and Fig. 3.12). It has been perceived that the development blocks with higher percentage of literates normally have better educational facilities, comparatively better economic condition and low proportion of SC population. The high percentage of literacy is also attributed to the better transport linkages to Varanasi city having large number of educational institutions. Taking the literacy sex-wise, male literacy dominates everything which is common feature throughout the country.

Female literacy (23.21 %) is much lower than the male literacy (62.76 %). As regards the development block-wise distribution, highest male literacy (above 80 %) is observed in Baragaon and Pindra development blocks while Harhua, Kashi Vidyapith, Araziline, Cholapur and Sewapuri development block have comparatively lowest figure (75–80%) and Chiraigaon blocks have literacy rate below 75 % (Fig. 3.13). Analysis of female literacy shows that it is higher (above 47 %) in Baragaon and Kashi Vidyapith development block. Araziline development block registers comparatively lowest (below 44 %) literacy rate because of less number of school and other higher educational centres particularly for females (Fig. 3.14). Parents do not permit their girl children to travel long distances for education rather they force them to engage in domestic and other agricultural works though the changing trend is apparent in new generation, which may yield fruitful results in future. Growth rate of literacy of each development blocks is shown in the Table 3.9.

3.7 Occupational Structure

The participation ratio of various working categories is one of the best indicators of the existing state of employment opportunities for the population of a region. The distribution and density of population have no noteworthy values unless and until the population pressure is examined in context of its aspect (Garnier 1978). The demographic, economic and social factors which are associated with the size and structure of population, determine the proportion of the workers in the region. Thus, occupational composition is an important index to through light upon the economic dynamic of health and vigour of a region. The working and non-working forces are the basic determines for the economic structure of human population in a region (Stamp 1957). A variety of economic, social and demographic factors are responsible to determine the magnitude of working force. The pressure on land in Varanasi

Table 3.8 Literacy rate and decadal difference in rural literacy rates by sex, Varanasi district, 1991–2001

Development block	1981			1991			2001			Decadal difference: 1991–2001 (%)		
	Male	Female	Total	Male	Female	Total	Male	Female	Total	Male	Female	Total
Baragaon	48.92	12.09	31.15	69.0	26.3	48.0	80.87	47.04	63.94	11.87	20.74	15.94
Pindra	47.62	11.57	29.77	68.0	25.4	47.1	80.42	45.97	63.28	12.42	20.57	16.18
Cholapur	43.48	4.93	28.13	61.6	23.2	43.2	75.36	45.11	60.62	13.76	21.91	17.42
Chirai Gaon	38.57	9.12	24.52	54.1	18.10	37.3	74.66	44.67	60.48	20.56	26.57	23.18
Harhua	43.57	9.61	27.6	62.5	22.6	43.6	77.48	46.52	62.77	14.98	23.92	19.17
Sewapuri	44.21	10.51	28.45	64.4	23.9	45.0	77.33	44.3	61.39	12.93	20.4	16.39
Araziline	42.82	8.45	27.03	62.3	21.4	43.20	75.21	43.15	60.08	12.91	21.75	16.88
Kashi Vidhyapith	41.68	8.47	26.46	60.2	23.9	43.50	76.44	48.46	63.45	16.24	24.56	19.95
Total Urban	56.2	36.6	54.4	64.8	48.77	64.77	79.87	63.81	71.96	15.07	15.04	7.19
Varanasi District	45.10	10.96	28.68	62.76	23.10	43.80	77.87	53.05	66.12	15.11	29.95	22.32

Source: Census of India, Primary Census Abstract (PCA), Varanasi district, 1991 and 2001 and self computed

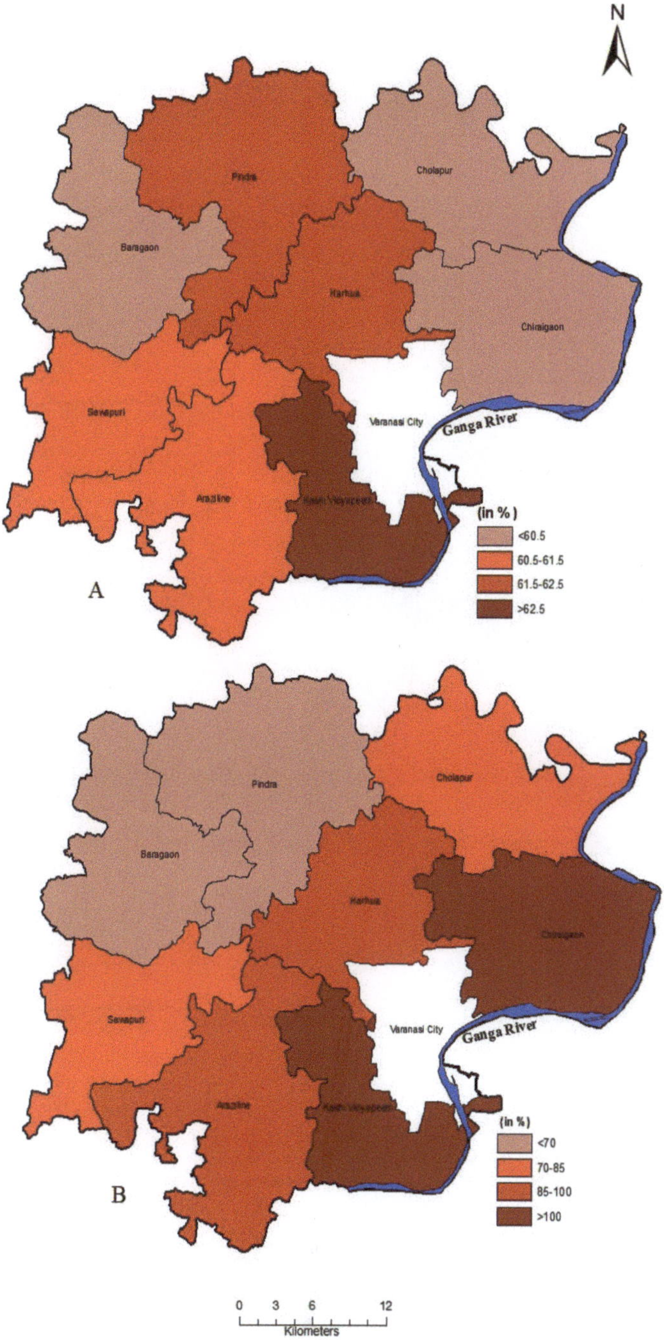

Fig. 3.12 Growth rate of literacy in rural area of each development blocks of Varanasi district in 1981–1991 (**a**) and 1991–2001 (**b**) respectively

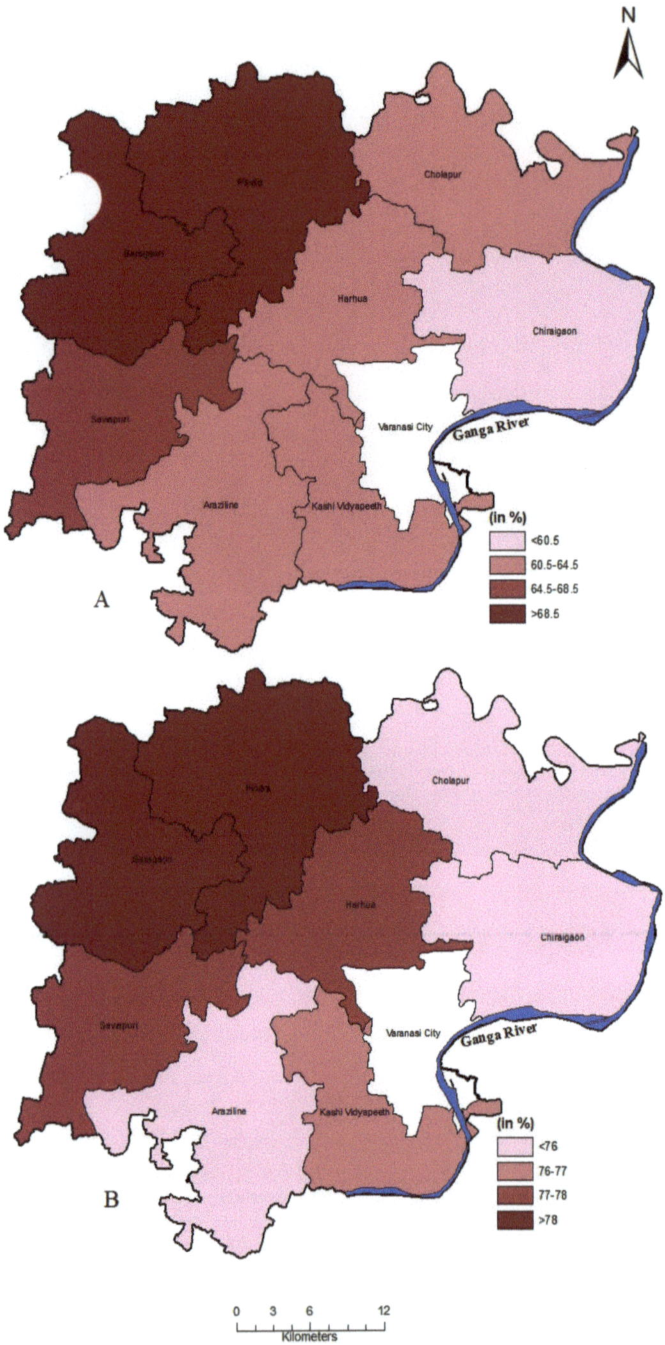

Fig. 3.13 Male literacy rate in rural area of each development blocks of Varanasi district in 1991 (**a**) and 2001 (**b**) respectively

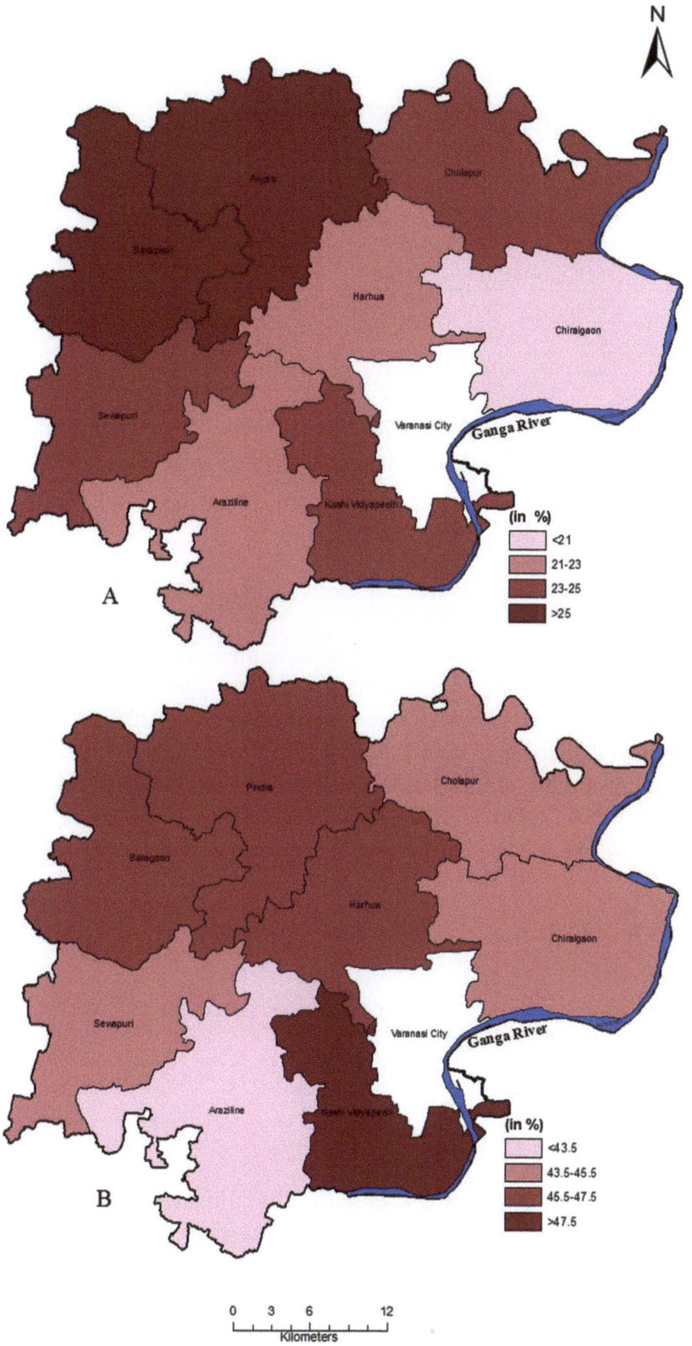

Fig. 3.14 Female literacy rate in rural area of each development blocks of Varanasi district in 1991 (**a**) and 2001 (**b**) respectively

Table 3.9 Growth rate of literacy in Varanasi district

Development block	1971	1981	1991	2001	Growth rate (%) 1971–1981	1981–1991	1991–2001
Baragaon	4004	40418	59978	99141	909.44	48.39	65.3
Pindra	4985	44428	71804	119950	791.23	61.62	67.05
Cholapur	4378	35463	54732	98423	710.03	54.34	79.83
Chirai Gaon	5528	34309	52949	119598	520.64	54.33	125.87
Harhua	4341	34385	55661	108411	692.1	61.88	94.77
Sewapuri	4085	34299	55112	99645	739.63	60.68	80.8
Araziline	6220	48104	77517	147117	673.38	61.14	89.79
Kashi Vidhyapith	5906	34046	62224	131354	476.46	82.76	111.1
Total District	39447	305452	489977	923639	674.34	60.41	88.51

Source: Census of India, Primary Census Abstract (PCA), Varanasi district, 1991 and 2001 and self computed

has mitigated because a sizable work-force was found to be engaged in secondary and tertiary sector.

The analysis of occupational structure reflects the nature of economy prevailing in a region. Agriculture is predominant of the rural part of study area engaging about 28.27 % of the total working force under it. In non-farming pursuits, 23.5 % people are engaged in house hold industries while above 36.27 % in other occupation. Out of the total population nearly 68.71 % are non-workers and mainly dependent on the working force (Table 3.10 and Fig. 3.15).

3.7.1 *Main Workers*

Workers who had worked for the major part of the reference period (i.e. 6 months or more) are termed as Main Workers.

3.7.1.1 Cultivators

For the purpose of census, a person is classified as cultivators if he or she is engaged in cultivation of land owned or held form government or held from private persons or institution for payment in cash or kind. Similarly, a person working on another person's land for salaries in cash or kind or both is treated as cultivator. The percentage of cultivators is quite high as compared to other category of workers. There were 34.08 % cultivators for the district as a whole in 1981 which decreased to 25.19 % in 1991 due to lack of interest in cultivation. The proportion of cultivators

Table 3.10 Occupational structure of Varanasi district, 2001 (in %)

Development block	Year	Cultivator	Agricultural labor	Household industry	Other services	Total main worker	Marginal worker
Baragaon	1981	50.39	13.87	11.52	18.09	93.87	6.13
	1991	43.57	11.49	11.56	5.08	85.33	14.67
	2001	35.89	7.12	13.35	17.96	74.02	25.97
Pindra	1981	60.12	14.52	7.71	17.81	98.42	1.58
	1991	53.30	13.52	4.50	6.41	91.20	8.80
	2001	31.47	5.50	6.21	19.30	62.50	37.49
Cholapur	1981	51.47	20.36	6.33	16.53	96.55	3.45
	1991	43.83	20.07	10.63	5.13	89.62	10.38
	2001	28.06	7.85	12.63	23.82	72.37	27.62
Chirai Gaon	1981	42.57	15.64	12.68	27.60	98.44	1.56
	1991	37.31	14.61	11.98	7.58	90.74	9.26
	2001	21.44	4.85	12.17	36.23	74.70	25.29
Harhua	1981	46.01	11.86	8.46	31.07	97.41	2.59
	1991	36.75	12.73	6.97	7.09	84.83	15.17
	2001	18.98	7.11	8.56	33.43	68.09	31.90
Sewapuri	1981	50.05	12.86	10.51	23.61	97.03	2.97
	1991	39.28	16.50	16.70	5.50	91.54	8.46
	2001	31.33	9.47	16.48	21.55	74.59	25.40
Araziline	1981	43.08	12.29	16.29	24.19	95.85	4.15
	1991	34.84	12.17	18.09	6.13	86.53	13.47
	2001	26.08	5.23	18.38	26.65	76.96	23.03
Kashi Vidhyapith	1981	35.52	9.40	13.79	38.30	97.01	2.99
	1991	26.08	5.23	18.38	26.65	76.96	23.03
	2001	13.05	3.29	17.11	42.94	76.41	23.58
Total Rural	1981	44.30	20.88	11.65	18.47	95.31	4.69
	1991	39.17	13.78	12.28	6.82	88.68	11.32
	2001	28.37	11.77	16.34	8.20	79.98	17.8
Total Urban	1981	2.88	2.16	21.51	73.03	99.59	0.41
	1991	1.58	1.02	24.38	20.00	98.31	1.69
	2001	5.12	0.98	28.7	–	96.6	–
Total District	1981	34.08	16.26	14.08	31.94	96.37	3.63
	1991	25.19	9.04	16.78	31.12	92.26	7.74
	2001	20.21	6.3	23.5	36.27	74.10	8.91

Source: Census of India, Primary Census Abstract (PCA), Varanasi district, 2001 and self computed

20.21 % also declined in 2001. Baragaon and Pindra block recorded highest proportion of cultivators at 35.89 % and 31.47 % in 2001 respectively. The lowest figure was found to be 13.05 % in Kashi Vidyapith development block in 2001 (Table 3.10 and Fig. 3.15).

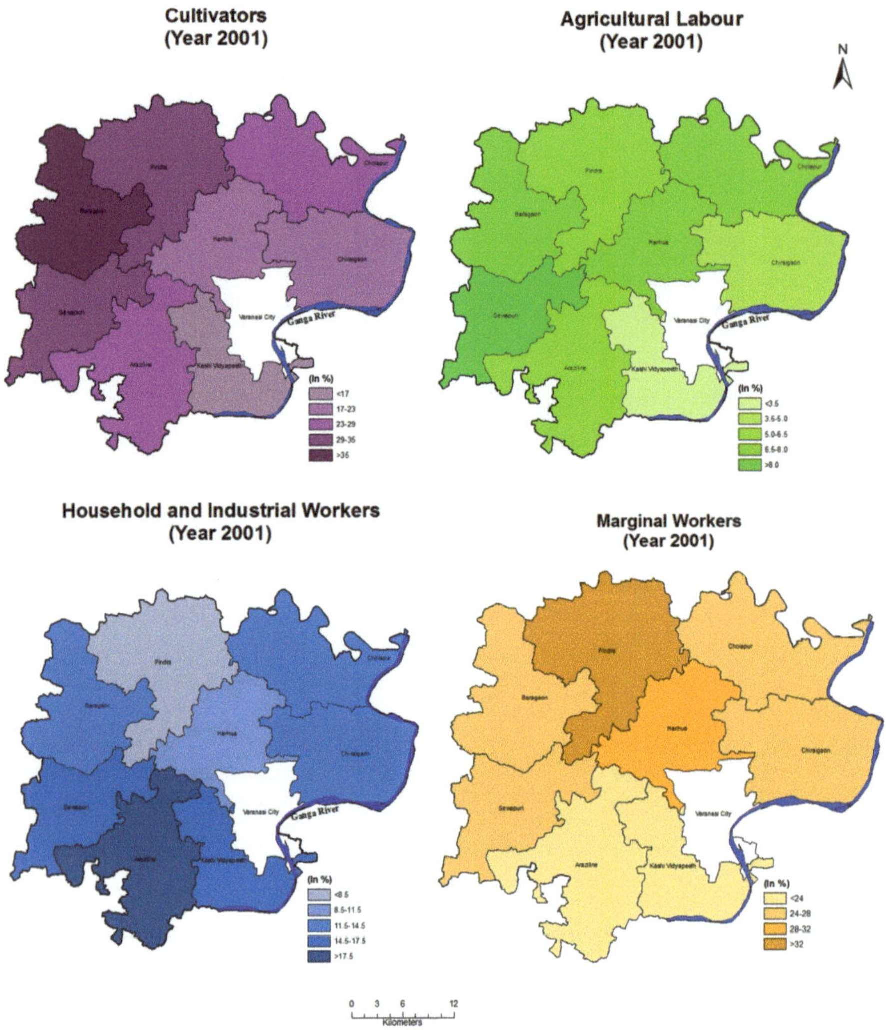

Fig. 3.15 Occupational structure in Varanasi district

3.7.1.2 Agricultural Labors

A person who works on another person's land for wages in money or kind or share is regarded as an agricultural labor. An agricultural labourer has no right of lease or contract on land on which he/she works. In 2001, proportion of agricultural laborers constituted of 6.3 % which was 16.26 % in year 1981 and ranked last in the main worker category. According to 2001 census the highest proportion of agricultural labourers was found in Sewapuri block (9.47 %) but it was highest at 20.36 % and 20.07 % in Cholapur block in 1981 and 1991 respectively. The lowest proportion of agricultural labour was found in Kashi Vidyapith block (3.29 %) followed by

Chiraigaon (4.85 %) and Araziline (5.23 %) respectively. The proportion of agricultural workers was highest among females than males in all the development blocks and in all the decades whereas highest proportion of cultivators was registered in rural area than in urban areas in all the decades (Table 3.10 and Fig. 3.15).

3.7.1.3 Household Industrial Workers

Household industry is well-defined as an industry conducted by one or more members of the household at home or within the village in rural areas and only within the limits of the house in urban areas. The share of such workers was 14.08 % in 1981, 16.78 % in 1991 and 23.5 % in 2001. It happened due to development of some agro-based industries in rural areas.

There is found wide variation in distribution of house industrial workers at block level. Pindra (6.21 %) and Harhua (8.56 %) have low-percentage of household's industrial workers in 2001. The medium percentage of this category (11–14 %) has been recorded in Chiraigaon (12.17 %), Cholapur (12.63 %) and Baragaon (13.35 %) blocks where households industries are few in numbers. The highest percentage of household's industrial workers (above 16 %) is recorded in Sewapuri, Kashi Vidyapith and Araziline blocks (Fig. 3.15).

3.7.1.4 Marginal Workers

Those workers who had not worked for the major part of the reference periods (i.e. less than 6 months) are termed as marginal workers. The percentage share of marginal working population to the total population in the district is 8.91 %. It is highest (37.49 %) in Pindra and lowest (23.03 %) in Araziline Block. Marginal working population in Harhua and Baragaon development blocks was 31.90 % and 25.97 % in 2001 respectively (Fig. 3.15).

3.7.1.5 Other Services/Workers

All the workers, i.e. those who have been engaged in some economic activities during last 1 year, but are not cultivators or agricultural laborers or in household industry, are 'Other Services/Workers'. The type of workers that come under this category include all government servants, municipal employees, teachers, factory workers, those engaged in transport, commerce, banking, mining, construction, political and social works etc. The district recorded 26.27 % of its total working force in other services in 2001, which was 36.27 % in 1991. There is good improvement in other service category. The increasing percentage of other workers shows the inability of agriculture to provide jobs for more hands and so people preferred to go out for other works than agriculture. It is observed that the percentage share of workers involved in other services in Pindra and Baragaon blocks was low (below 20 %)

while it is 21–30% in Araziline, Sewapuri and Cholapur blocks in 2001. Kashi Vidyapith, Chiraigaon and Harahua blocks records 42.94 %, 36.23 % and 33.43 % of their workers in other services respectively (Table 3.10). The proportion of the other workers was very high in urban areas than in rural area in the district.

3.7.2 *Non-Workers*

Persons who did not work at all during the reference period were treated as non-workers. The proportion of non-workers to the total population in the Varanasi district was 5.42 %.

References

Census of India, 2001.

Chandrashekhar, S. (1967). *India's population facts problems and policy* (p. 79). Bombay: Allied Publishers Pvt. Ltd.

Chandana, R. C., & Sidhu, M. S. (1980). *Introduction to population geography* (pp. 79–81). New Delhi: Kalyanai Publication.

Chandna, R. C. (1986). *Geography of population – Concept, determinants and patterns* (p. 100). New Delhi: Kalyani Publishers.

Clark, J. I. (1977). *Population geography* (p. 258). New York: Pergaman Press Ltd.

Demco, G. T., Rose, H. M. & Schnell, C. A. (1970). Population distribution and density: Spatial characteristics and changing pattern. In A. Reader (Ed.), *Population geography* (Part IV, p. 154). New York: McGraw Hill Book Co..

Ezimmermann, E. W. (1951). *World resource and introduction* (p. 86). New York: Harper Row.

Franklin, S. H. (1956). The pattern of sex-ratio in New Zealand. *Economic Geography, 32*, 168.

Garnier, J. B. (1978). *Geography of population* (p. 33). (S. H. Beaver, Trans.). London: Longman.

Ghosh, B. N. (1985). *Fundamental of population geography* (p. 20). New Delhi: Sterling Publishers Pvt. Ltd.

Ghosh, B. N., & Mukharjee, A. B. (1989). An analysis of health services coverage of primary health centre in West Bengal. *Indian Journal of Public Health, 33*(1), 12–19.

Kinsley, D. (1959). The sociology of demographic behavior. In R. K. Merton, L. Broom, & L. S. Cottsell Jr. (Eds.), *Sociology today* (p. 314). New York: Basic Book Inc.

Sinha, V. C., & Zacharia, E. (1979). *Elements of demography* (p. 482). New Delhi: Allied Publishers.

Stamp, L. D. (1957). *Asia: A regional and economic geography* (p. 254). London: Methuen Ltd.

Thomlinson, R. (1965). *Population dynamic* (p. 428). New York: Random House.

Trewartha, G. T. (1953). A case for population geography. *Annals of the Association of American Geographers, 43*(2), 71–97.

Varma, K. N. (1967). *Population problem in the Ganges Valley* (1st ed., p. 37). Agra: Shiva Lai Agarwala.

Chapter 4
Analysis of Health Care Facility Using GIS and GPS

Abstract Analyzing distribution of hospitals through GIS and GPS is a significant measure in health care facility because every category of population should get access to the hospital facility optimally. Spatial analysis of health care facilities using GIS is analysed in this chapter. Health care service in Varanasi district go from bad to worse, all it shortage of man power, especially of doctors, or negligence in providing health services, many government hospitals, primary health centres (PHCs), community health centres (CHCs) seem to be witnessing health care delivery services at their worst in the region. Buffer (Proximity) analysis is used widely for many situations – e.g. to understand the association between transportation facilities in the study area to existing health care facilities. Buffer technique also play a vital role in the health GIS application through which we can easily calculate the number of persons live within a 10 km. radius from a particular primary health centre's (PHCs) or community health centres (CHCs) or from the other governmental hospitals etc. of the district. By applying the proximity analysis of health centres, it is found that maximum rural population are totally depended on the existing government health centres. Shortest route estimation through network analysis is used for identifying the most efficient routes or paths for allocation of services.

Keywords GIS • GPS • Primary health centres (PHC) • Community health centres (CHC) • Health care service • Buffer analysis • Network analysis

4.1 Health Care Facilities

Analyzing distribution of hospitals is an important criterion in health care facility because every category of population should get access to the hospital facility optimally. Spatial distribution of health care faculties is most effective factor in assuming the utilization pattern of health care facilities in an area (Aday and Andersen 1974). Further, uneven distribution of medical facilities is a basic problem in utilizing health services in a rational manner. Wide varieties of problem arise in utilizing such facilities on account of spatial variations and Varanasi is not exception of this characteristics.

P.K. Rai, M.S. Nathawat, *Geoinformatics in Health Facility Analysis*,
DOI 10.1007/978-3-319-44624-0_4

Health care service in Varanasi district go from bad to worse, all it shortage of man power, especially of doctors, or negligence in providing health services, many government hospitals, primary health centres (PHCs), community health centres (CHCs) seem to be witnessing health care delivery services at their worst in the region. A number of loopholes in health care services at the hospitals, poor sanitary conditions around hospitals premises also raise eyebrows. Thus, it has become important to recognize local healthcare needs given the healthcare demands, social status and services that are available in the locality and geographic information system is very helpful for the same (Rai et al. 2011).

In order to analyse the utilization pattern of health care facilities in any spatial unit, it is pertinent to first assess the distribution of health care facilities in spatial perspective. The member country of WHO adopted a resolution in 1978 to achieve the goal of 'health for all by 2000 AD'. In view of the above objective, the Government of India has proposed to expand and provide health centre on every 30,000 population in plains (on 20,000 population in hilly and tribal area), a sub-centre on every 5000 population in plain area (on 3000 population in hilly and tribal areas) and a CHC on every 120,000 population in plain area (80,000 population in hilly and tribal area). In the present context it has become essential to analyse the availability of health care facilities in the study area, according to suggested norms of country's health care policy. It will help in providing proper health services to the rural population and framing plan for rational distribution of health care services to the common people (Rai et al. 2011).

The geographical information system (GIS) is very helpful in a variety of application areas points to an increasing interest in the spatial aspects of health policies and planning. The increasing availability of Geographical Information Systems (GIS) in health organizations, together with the proliferation of spatially disaggregate data, has led to a number of studies that have been concerned with developing measures of access to health care services (Rai et al. 2011). This can be effectively used as a tool for decision making in relation to optimum and gainful utilization of available medical resources in any areas. Health care facility and utilization is concerned with all the issues that are related to the locations and facilities. These issues include the optimal location of PHCs, CHCs, hospitals and clinics, the relationship between existing locations and health care needs and assessment of hospitals and the assessment of facilities. Analyzing distribution of hospitals is an important criterion in health care facility because every category of population should get access to the hospital facility optimally (Rai et al. 2011). Keeping this view author mainly discussed existing health care facilities in the study area. Latest technologies like GIS and GPS are used to locate the existing health care facilities in the study area.

4.2 Spatial Analysis of Social Indicator in Varanasi District

Since health services are an essential part of social welfare, it is necessary to analyze the variation in social indicators which directly influence to health status of the people and required facilities of health care. Analysis of data collected for 8 development blocks reveals inter and intra-regional variations at block level for different social indicators. Here social indicators include densities of settlement, household and population per $km^{2.}$ The density of settlement is found to be highest in Harahua block (1.2/km^2) followed by Sewapuri (1.11/km^2), Araziline (1.0/ km^2) and Kashi Vidyapith (0.86/ km^2). The lowest density of settlement per km^2 has been observed in Chiraigaon and Baragaon blocks as 0.71 and 0.78 respectively.

In terms of densities of households and population, Kashi Vidyapith ranks first incompassing the values 173.0/ km^2 and 1266/km^2 respectively. In this context the blocks such as Harhua and Araziline posses higher density of both aspects than Pindra, Cholapur and Baragaon blocks. Minimum density of household (116 household/km^2) is found in Sewapuri development block whereas lowest population density (906/km^2) calculated for Baragaon development (Table 4.2). The above analysis clearly state that the both household and population densities in most of the blocks are quite high and therefore there is need of effective health care system in the study area.

4.3 Spatial Analysis of Health Care Facilities in Varanasi District

Health facilities of Varanasi district are based on mainly modern allopathic of treatment. To know the distributional pattern of health care facilities, data has been collected from Chief Medical Officer (CMO) office and government hospital/health units located in rural areas of Varanasi district. There are different categories of health centre providing infrastructure and treatment in the district. The PHCs are dotted in the district located at an interval of 10–20 kms and the tahsil hospitals are located about 50 km apart.

The hierarchical with the distribution of medical centres of the district bears a close relationship with the hierarchy a central places and population size of the settlement (Rai et al. 2011).

Besides, the transport network has also influenced the growth of health care facilities. Figure 4.1 and 4.2 show the spatial distribution of government health care units in eight blocks of Varanasi district in the year 2001 and 2009 respectively. Table 4.1 shows the block wise distribution of health care facilities, namely Aushadhalayas, PHCs, proposed/existing Sub-Centres, Ayurvedic, Homeopathic and Unani health centres during year 2008–2009.

It is clear from the Table 4.1 that Aushadhalaya are mostly confined to urban centres besides one each in Baragaon, Cholapur, Chiraiagon, Harhua and Sewapuri development blocks respectively.

Out of total 26 Ayurvedic hospitals in the district the maximum number (5) is found in Harhua block followed by Cholapur (4), Kashi Vidyapith (3 each) except Baragaon (2). Homeopathic hospitals are confined mainly in Cholapur (3), Chiraigaon (3), Pindra (2) and Kashi Vidyapith (2). One Unani hospital situated in Kashi Vidyapith block of the study area. Since people living in rural areas, therefore, it is pertinent to analyse their distribution pattern in the district. Presently, the district as a whole possesses 32 PHCs (8 old PHCs and 24 new PHCs) and 304 sub-centres. In addition, as many as 64 referral centres (eight centres in each block) are being made operational for providing mother and child health care. Against the norms of one sub-centres for every 5000 rural population, a sub-centres in the district caters the health needs of over 8000 rural populations. Most of the sub-centres are being run by one ANM due to shortage of basic health workers. Shortage of health staff has also stopped these centres from carrying out routine immunization programme (twice a week). The analysis of available information reveals that considerable amount of these facilities are found in Araziline, Chiraigaon, Baragaon and Cholapur development blocks while Harhua, Pindra, Sewapuri and Kashi Vidyapith blocks are endowed with poor status in health care facilities (Rai et al. 2011).

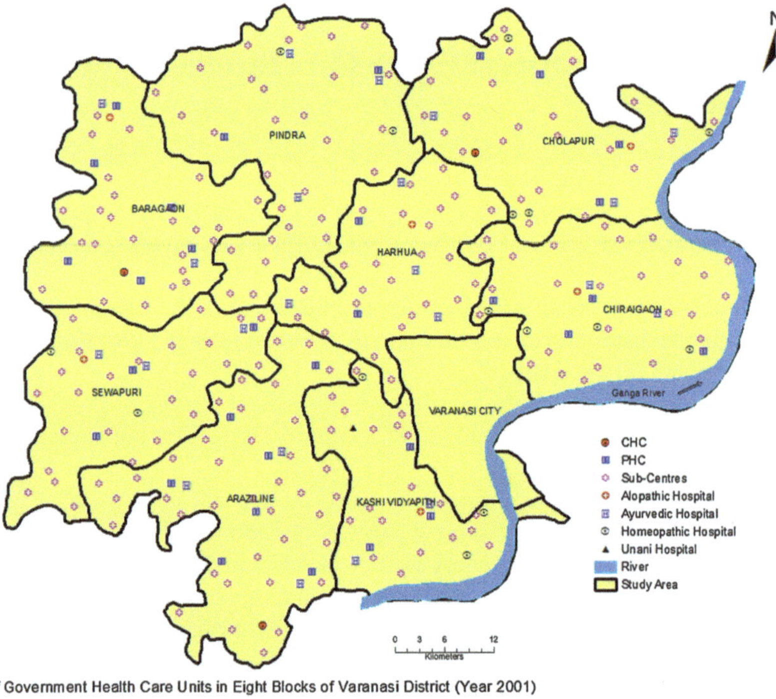

Fig. 4.1 Distribution of health care units in Varanasi district, 2001

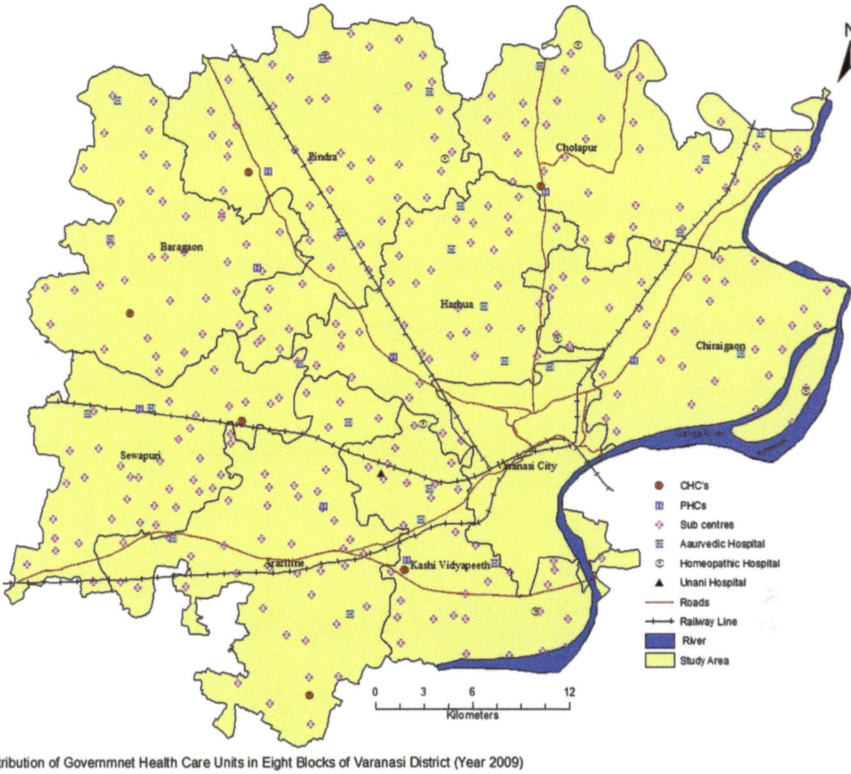

Fig. 4.2 Distribution of health care units in Varanasi district, 2009

The local people in the area travel a long distance to reach the PHCs for addressing their health problems, the sub-centres has also failed to provide mother and child health care and family planning, as it depends on block PHCs for most of health services (Rai et al. 2011).

Some of the rural belts in the study area including Padao, Ramnagar and Lohta are in the grip of quacks, who do not mind playing with the lives of innocent villagers. Due to lack of awareness and knowledge, even the residents of urban areas at times fall prey to these quacks. Around more than dozen unauthorized nursing homes and many privately operated pathological centres with blood bank facilities is running smoothly since many year but all are closed in the past 1 year. These health centres and some maternity clinic are found to be operating near government hospitals in the city and most of them are not registered (Rai et al. 2011).

Table 4.1 Distribution of government health care units in Varanasi district, 2009

S.No.	Development block	Ayurvedic hospital	Block wise PHC	CHC	Proposed/new PHC	No. of sub centres	Homeopathic hospital	FWC	MCWC	Aushada-laya
1.	Baragaon (204803)	1. Kathiraon	Baragaon	Birawkot	1. Dandupur	36	–	38	36	1
					2. Devchandrapur					
		2. Anei			3. Birawkot					
					4. Barhi Newda					
2.	Pindra (247525)	1. Mangari	Pindra	Pindra	1. Kashipur	47	2	49	47	–
		2. Sindora								
		3. Hiramanpur								
3.	Cholapur (205251)	1. Rajbadi	Cholapur	Cholapur	1. Niyardih	34	3	36	34	2
		2. Dahurahra			2. Danganj					
		3. Chubepur			3. Dahurahra					
		4. Chitaipur								
4.	Chiraigaon (234400)	1. Vraahi	Chirai Gaon	–	1. Sarnath	34	3	38	36	1
		2. Jalhupur			2. Gobaraha					
		3. Ranipur Khurd			3. Chhitauni					
					4. Kadipur					
5.	Harhua (210582)	1. Lalpur	Harhua	–	1. Puarikala	36	–	38	36	1
		2. Palhipatti								
		3. Ayar								
		4. Kajisarai								
		5. Lamhi								
6.	Sewapuri (200678)	1. Sewapuri	Sewapuri	Hathi Bazar	1. Pachvaar	34	–	35	33	1
		2. Rameswar			2. Domela (Baghvanpur)					
		3. Kapsethi								

7.	Araziline (297770)	1. Mirza Murad	Araziline	Araziline	1. Gangapur	44	–	46	44	–
					2. Mirzamurad					
		2. Bhadradri			3. Mahgaon					
					4. Payagpur					
		3. Kaparphorva			5. Rajatalab					
					6. Jagardevpur					
8.	Kashi Vidyapith (233676)	1. Rohaniya	Kashi Vidyapith	Misirpur	1. Ramna (2)	39	2	41	39	1
		2. Chitaipur								
		3. Lohta								
Total block (3138671)		26	08	06	24	304	10	321	305	7
Total urban (1312342)		152	–	–	–	–	–	–	–	–

Source: CMO Office, Varanasi District and Self Computed Rai et al. (2011)

4.3.1 Density of Health Care Facilities

It is a vital indicator of state of health care facilities in an area. The size (area) of block varies notably in the study area as well in the state to analyse health care facilities in terms of per 100 km^2 area has been taken into consideration. The density of medical institution per 100 km^2 ranges between 14.3 in Pindra to 26.7 in Kashi Vidyapith block followed by Araziline, Sewapuri, Harhua, Cholapur and Chiraigaon (Table 4.2). The establishments of New PHCs, sub-centres (linked to block PHCs) has turned out to be a useful affairs as it has failed to provide solutions to the basic health need to the local people. The study area has already witnessed establishment of as many as 306 sub-centres. Similarly, as many as 337 ANMs are also deployed in various sub-centres and other health centres for carrying out mother and child health care. Figure 4.3a–d to show density of medical institution/100sq. km., vvdensity of doctors/100sq.km, density of PHCs and CHCs Beds/100sq. km and density of paramedical employee/100sq. km in the year 2009 respectively

4.3.2 Distribution and Condition of Bed in the PHCs and CHCs/Hospitals

Another important indicator of sate of health care facility is considered as number of beds available in PHCs per 100 km^2 area. Density of beds available in PHCs varies between 5.6 beds/100 km^2 area in Harhua block and 27.97 bed/100 km^2 in Baragaon block. The high bed density is found in between 20 and 28 beds/100 km^2 in four blocks namely, Baragaon, Araziline, Kashi Vidyapith and Cholapur where as low beds density (10–20 beds/100 km^2) is found in Chiraigaon and Pindra block. Table 4.2 reveals that very low bed density (<11 beds/100 km^2) is recorded in Harhua and Chirai Gaon block of the district.

The analysis of distribution of beds in PHCs reveals that number of blocks have still lower density of beds than the standard suggested under National Health Policy. The worst condition are always witnessed in general wards of the government hospital are several beds are found without bed sheets and mattresses. In several time and in several places in the hospitals, mattress on beds are toiled with bloods stains and holes.

4.3.3 Availability of Doctors and Paramedical Staff

These are considered as the most dominating attribute of availability of health care facility in an area. The shortage of health staffs like doctors, paramedical employees, lab technician etc. at PHCs and CHCs are responsible for lack of adequate health care facility in the study area. The shorter of doctors is hampering the health

Table 4.2 Density of social indicators and health care facilities, 2009

		Density (per sq.km)			Density (per 100 sq. km.)			
	Development block -wise	Settlement	Households	Population density	PHC beds	Doctors	Paramedical employee	Medical Institution (M.I.)
1.	Baragaon	0.78	148.05	1096.65	27.97	7.27	28.54	19
2.	Pindra	0.85	147.28	1067.68	17	4.026	26.4	14.3
3.	Cholapur	0.82	158.25	1139.06	25.64	4.45	29.54	20.6
4.	Chiraigaon	0.71	175.84	1269.86	10.2	3.06	27.55	20.4
5.	Harhua	1.2	207.41	1517.11	5.6	2.1	32.98	21
6.	Sewapuri	1.1	158.81	1210.7	18.92	5.91	32.53	21.2
7.	Araziline	1.0	187.85	1424.04	26.5	5.93	30.15	26.4
8.	Kashi Vidhyapith	0.86	249.86	1763.28	26	2.05	35.59	26.7

Source: CMO Office, Varanasi District and Self Computed

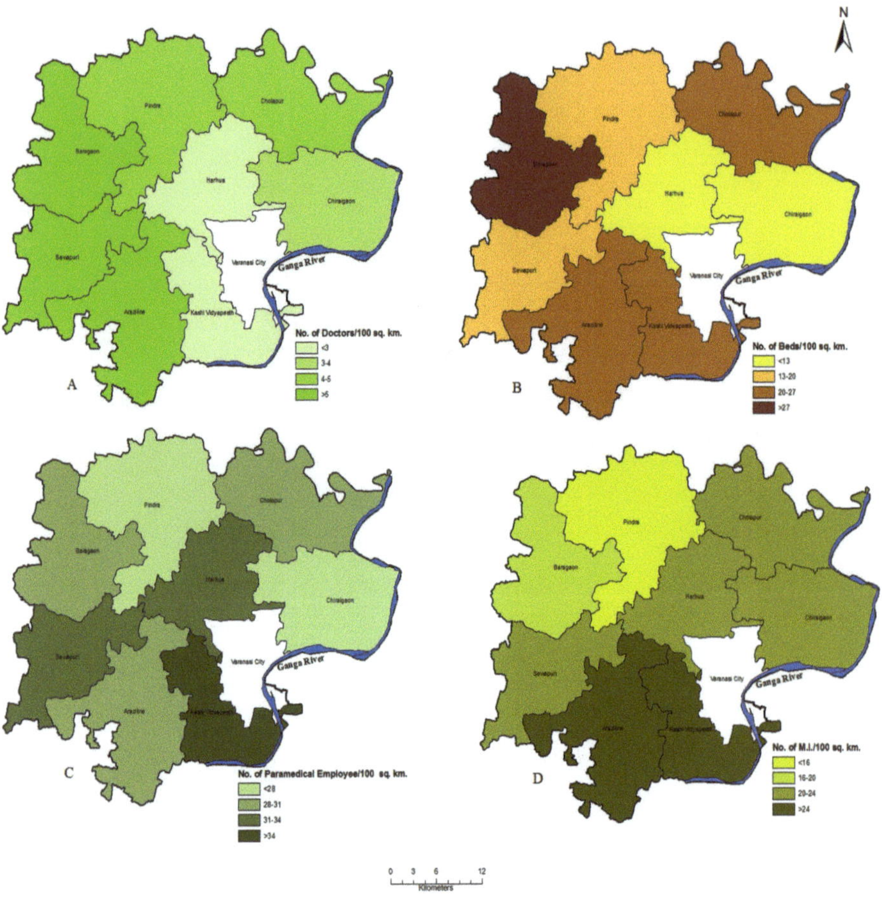

Fig. 4.3 (**a**) to (**d**) Density of health care facilities in 2009

care services, especially in the rural areas, as only 51 doctors are presently deployed at the PHCs and CHCs. There is a yawning gap between the sanctioned post (51 out of 130) of doctors and doctors actually deployed at these centres, the gap has been further widened in case of grass root health workers (multi-purpose workers), as only one-third (37) of the sanctioned post (112) has been filled so far. The study area is also facing a shortage of doctors specializing in gynecology and obstetrics that has crippled MCH service both at the CHCs and the government women hospitals. Only 11 women doctors are deployed against 19 sanctioned posts at government women hospitals while only four out of six CHCs in the district have gynecologists. The highest density of doctors 100/km^2 area is found in Baragaon block (7.27) whereas the lowest density is recorded in Harhua and Kashi Vidyapith development blocks which are 2.1 and 2.5 respectively. Pindra, Cholapur, Chiraigaon, Sewapuri, and Araziline development blocks possess a density of 2–6 doctors/100 km^2.

It is apparent from the analysis that distribution of doctors in the study-area is inadequate and uneven. There are five blocks where very low density of doctors per 100 km^2 area is marked. It is quite low as compared to norms put by the government of India in providing health care facilities for all in rural areas of the country.

Table 4.2 comprising social indicators and health indicators reveal that the development blocks having high density of social indicators than the density of health care facilities.

4.3.4 Distribution of Health Care Facilities in Terms of Per Lakh Population

Since all the blocks of the study-area are not uniform in terms of population distribution so that the distribution of health care facilities e.g. number of medical institutions, doctors and beds per 100,000 population required special attention for the present study. Further, the analysis of data regarding availability of health care facilities for two different periods of time is quite useful in evaluating different in equalities in the spatial distribution of health services.

Table 4.3 clearly shows that the number of medical institutions per 100,000 populations during year 2001 is lowest in Kashi Vidyapith block (11.6) and highest in Baragaon block (16.3). The availability of medical institutions has slight increased during the period from 2001 to 2009 in each block.

Although with an increase in population, health care facilities should also increase in number yet the situation remains the same. For the district as a whole the number of medical institution per 100,000 populations has been worked out as 16.25 in the year 2009. Of the total number of development blocks in the district only three blocks i.e. Pindra, Harhua and Kashi Vidyapith are behind the district average. It is further observed that four blocks- Baragaon, Cholapur, Sewapuri and Araziline are characterized with higher number of medical institutions per 100,000 persons than the district average in the year 2008–2009 (Fig. 4.4a–b). The number of doctors/ 100,000 populations is 5.8 in 1995–1996 which has marginally been increased to 5.66 in 2001 and then this figure decreased drastically and reached to 3.54 in year 2009. There are five development blocks i.e Pindra, Cholapur, Chirai Gaon, Harhua, and Kashi Vidyapith which have below 4 doctors per 100,000 populations. Table 4.3 also reveals that condition of Kashi Vidyapith is not very good for the same; and shows only 1.1 doctors in per 100,000 populations in 2009 whereas Baragaon block shows maximum 6.6 doctors per 100,000 populations. The number of doctors per 100,000 populations has not also increased significantly during the last 1995–1996 to 2001 (Fig. 4.5a–b). The change in number of doctors per 100,000 populations for the study-area during 1995–1996 to 2001 is calculated as only 1.44 % which is insignificant looking into the requirement of the area at that time. From the above analysis, it is reflect that the no of doctors per 100,000 populations showed negative percentage change during 2001–2009.

Table 4.3 Availability of health care facility per lakh population, 2009

Development block		No. of medical institution/lakh population (2001)	No. of medical institution/lakh population (2008–2009)	No. of doctor/lakh population (1996)	No. of doctor/lakh population (2001)	No. of doctor/lakh population (2008–2009)	No. of beds/lakh population (2005)	No. of beds/lakh population (2001)	No.of beds/lakh population (2008–2009)	No. of CHCs & PHCs/lakh population		
										(2001)	(2006)	(2008–2009)
1.	Baragaon	16.3	17.3	8.7	8.7	6.6	38.9	30.1	25.5.	4.3	3.1	2.9
2.	Pindra	13.0	13.4	5.1	3.0	3.7	7.1	5.9	15.9	1.0	1.3	1.2
3.	Cholapur	15.6	18.1	8.6	8.6	3.9	29.6	23.5	22.5	3.1	2.4	2.4
4.	Chiraigaon	14.0	16.0	4.3	5.4	2.4	10.8	8.0	8.0	3.2	2.4	2.1
5.	Harhua	12.5	13.9	4.2	2.4	1.4	4.8	3.7	3.7	1.8	0.9	0.9
6.	Sewapuri	14.6	17.6	3.8	3.8	4.8	23.3	7.8	15.6	2.5	2.0	1.9
7.	Araziline	15.3	18.6	6.8	8.5	4.1	24.7	19.9	18.6	3.0	2.6	2.6
8.	Kashi Vidhyapith	11.6	15.1	3.2	4.9	1.1	20.5	14.8	14.7	1.6	1.2	1.7
Total		14.11	16.25	5.58	5.66	3.54	19.96	14.21	15.58	2.6	1.98	2.0

Source: CMO Office, Varanasi District and Self Computed Rai et al. (2011)

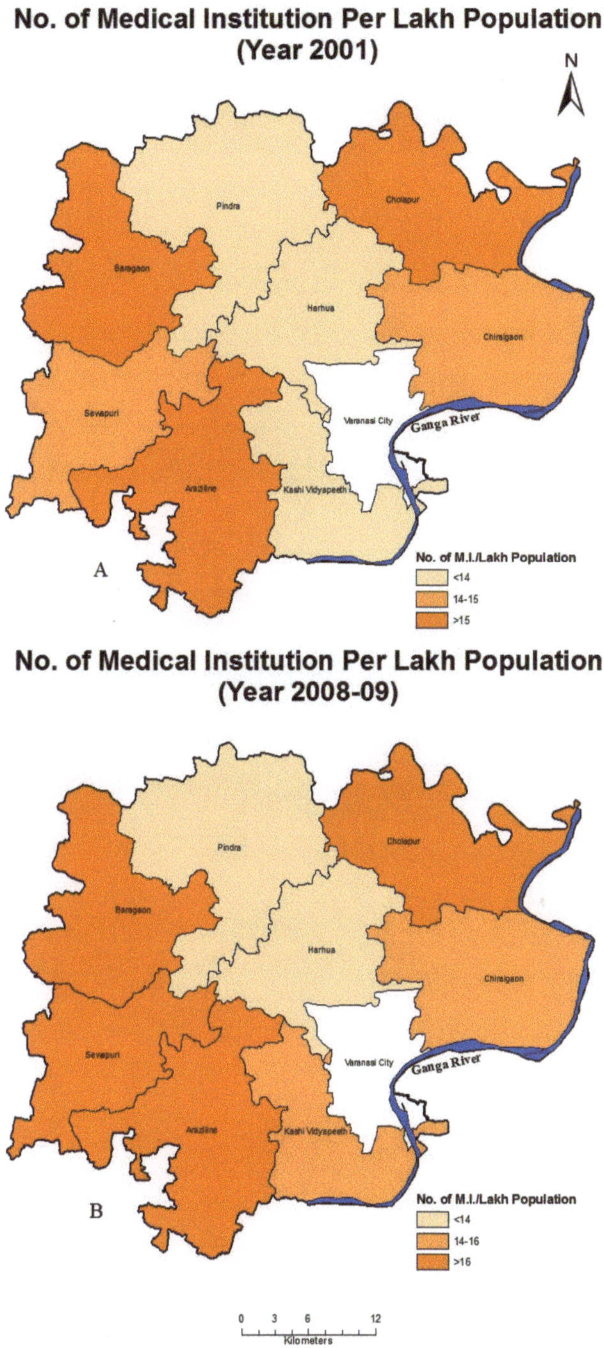

Fig. 4.4 (**a**) to (**b**) No. of medical institution per lakh population in 2001 and 2008–2009 respectively

Fig. 4.5 (**a**) to (**b**) No. of doctors per lakh population in 2001 and 2008–2009 respectively

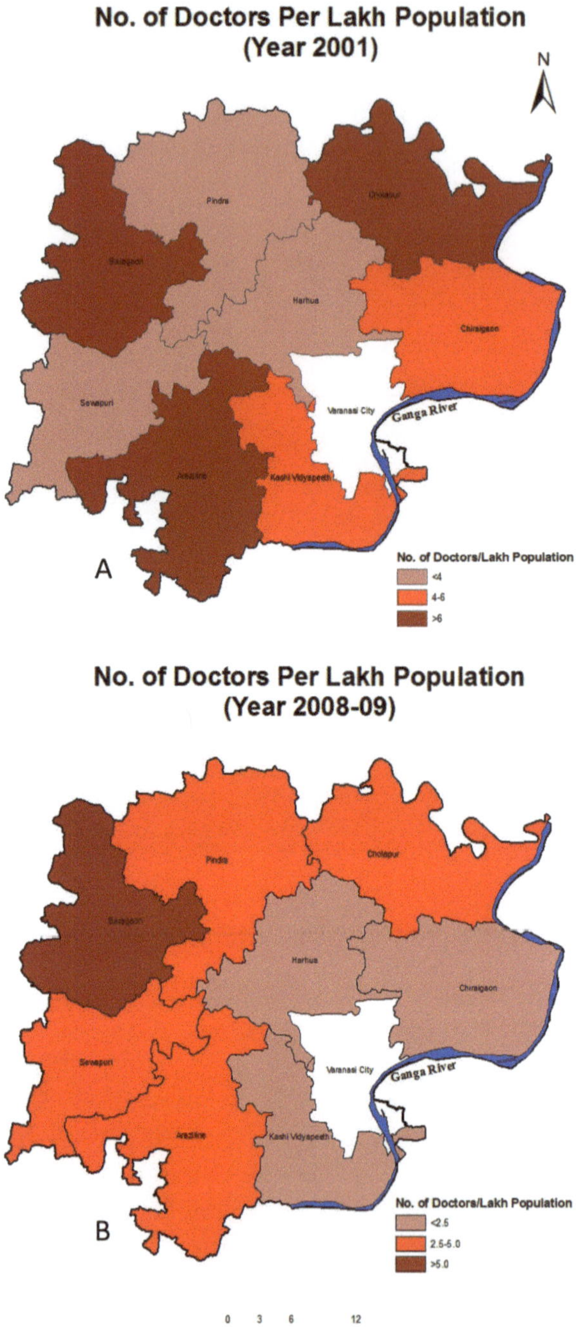

During these 15 years (from 1995–1996 to 2009), number of doctors per 100,000 populations in each blocks except Sewapuri showed negative percentage change. It means that the state government could not provide recommended number of doctors including the specialized doctors at PHCs of Varanasi district. It perhaps due to financial crisis and non-recruitment of doctors for PHCs. The unwillingness on the part of doctors in joining government health care centres in remote rural areas is also worsening the situation. The condition can be improved if increased incentives and facilities are provided to the doctors who are deployed in the rural areas. The health centres generally care for outdoor patients only. A few beds are provided for maternity cases or for patients who need observation out various types of primitive health programmes. Therefore, there is a need to analyze number of hospital bed available in the rural areas of the district.

It is further observed that the position of availability of beds in PHCs per 100,000 populations in 1995–1996 is better in blocks namely Baragaon (38.9), Cholapur (29.6), Araziline (24.7) and Sewapuri (23.3). The higher number of beds available per 100,000 populations in the block may be attributed to their smaller size of population as compared to the district (Fig. 4.6a–b). It is worth mentioning that no significant increase in number of beds in PHCs has taken place during 2001–2009 periods and it showed negative percentage change. Only Pindra block experienced positive percentage change in terms of number of beds per 100,000 populations due to rapid increase in population and rest of blocks showed negative percentage change. This whole study clearly reflects that during the last 10–15 years no significant change in state of government health care facilities in rural areas of Varanasi district. There is need to pay more consideration towards improving satisfactory health care facilities to the population living in rural areas of Varanasi district. Figure 4.7a–b shows the No. of PHCs and CHCs/Lakh Population.

4.4 Physical Status and Other Available Facilities at Primary Health Centres

The establishment of Primary Health Centres in India had started as early as in 1952 and till then it had undergone to significant changes to meet the ever increasing demand of health care services. Until the Eight Five Year Plan emphasis is mainly put on consolidation to the existing health qualitative rather than expansion. The main emphasis is given to improve qualitative nature of the health services through strengthening physical facilities like provision of essential equipment, supply of essential and life saving drugs, construction of building and staff quarters, filling up of vacant post of doctors and permanent staff.

Government of India in Bulletin on Rural Health Statistics in India (Ministry of Health and Family Welfare, New Delhi, 1998) highlights on the state of health care facilities available at different level such as at PHC in terms of infrastructure available in the study area. This analysis is basically based on rigorous survey of all the old PHCs (8 in number) of each development block and 24 newly developed PHCs

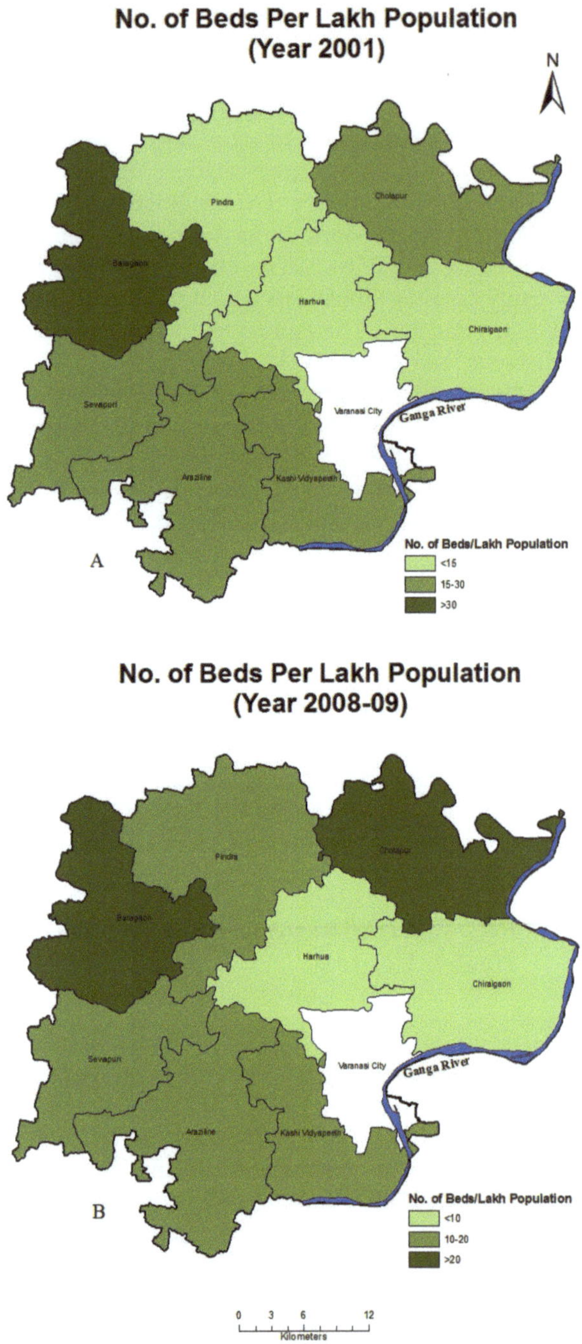

Fig. 4.6 (**a**) to (**b**) No. of beds per lakh population in 2001 and 2008–2009 respectively

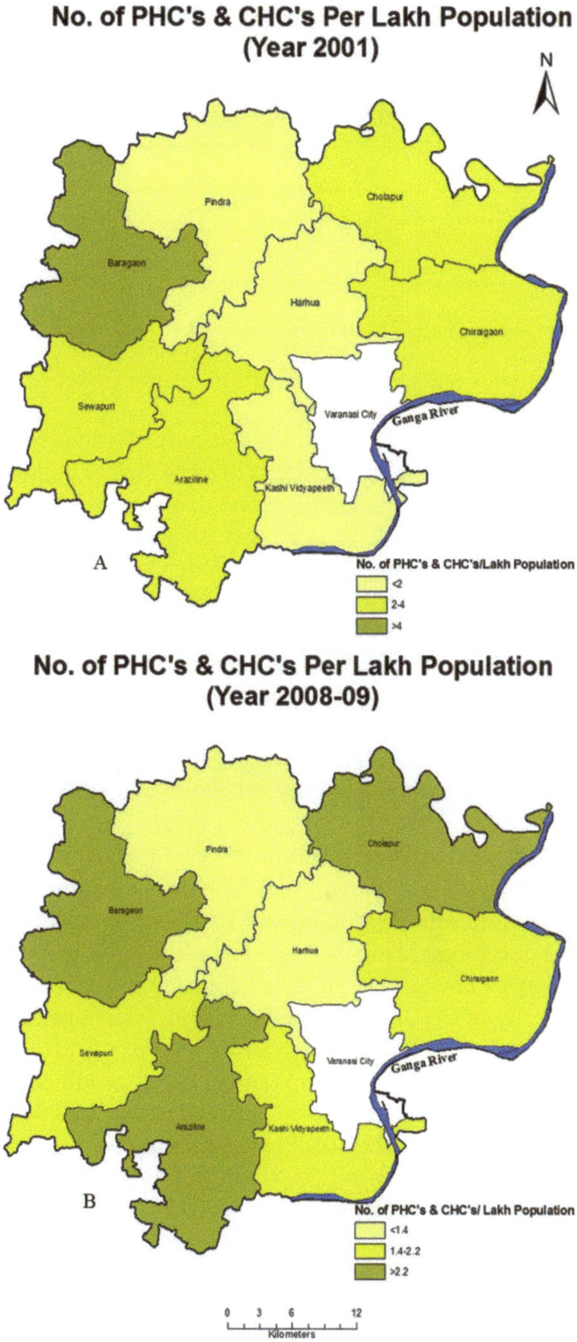

Fig. 4.7 (**a**) to (**b** No. of PHCs and CHCs per lakh population in 2001 and 2008–2009 respectively

Table 4.4 Block-wise percentage of PHCs having doctors in 1999 and 2009 respectively

Development block	No. of doctors (1999)	No. of doctors (2009)	Percentage (1999)	Percentage (2009)
Baragaon	14	13	77.77	72.20
Pindra	06	09	100.00	50.00
Cholapur	14	08	87.50	66.66
Chiraigaon	10	06	66.66	40.00
Harhua	04	03	66.66	50.00
Sewapuri	06	10	66.66	111.11
Araziline	20	13	95.23	61.90
Kashi Vidyapith	09	03	100.00	33.33
Total	83	65	86.45	67.70

in different blocks. The staffing pattern of primary health centres reflects that it is not adequate to fulfill the needs of large size of population of rural area (Table 4.4). As for disease concern, only one district malaria officer (DMO) is posted in the Varanasi District. A total of only 14 medical officers (8 in Block PHC and 6 in Community Health Centre) are working here for this large populated area. Likewise condition and strength of other medical employees are not sufficient to maintain here health facilities. Table 4.5 and 4.6 evince the facilities available at PHCs in different development blocks. It is assumed that if adequate infrastructure facilities in terms of quantity as well as quality are available, only then the utilization of health care facilities will be efficient in all sections of the society. In the case of PHCs infrastructure facilities include buildings of PHCs doctor's residence, water supply and availability of electricity, telephone and vehicles etc. (Rai et al. 2011).

In order to function the PHC, a building is required. The building may be of its own, rented, or any other type. Table 4.6 shows that on an average only 78.1 % of the PHCs of study-area have their own building. About 12.5 % and 9.3 % PHCs function in rented building and Panchayat Bhawan respectively. The condition of two development blocks, with respect to state of buildings namely Chiraigaon and Sewapuri is worst. In three development blocks (Baragaon, Chiraigaon, Sewapuri) majority of PHCs are in operation from Panchayat Bhawan only. It is further noted that 20 % of PHCs of Chiraigaon, Sewapuri, Araziline development block function in rented buildings. In terms of nature of buildings e.g. pucca (cemented) or semi pucca (semi-cemented), it is satisfactory to know that, all the PHCs of eight development blocks of the district are operating from pucca (cemented) building.

In order to make efficient utilization of health care facilities, 24 h presence of doctors at PHC is a pre-requisite. To ensure doctor's presence, doctor's residence must be available at each PHC. In this context it seems very disheartening that only 40.6 % of the total PHCs have provision of doctor's residence within PHCs premises. The availability of doctor's development block appears quite satisfactory. It is distinguished that PHCs located at Cholapur; Sewapuri and Kashi Vidyapith blocks

Table 4.5 No. of PHC (Block), PHC (Additional), Sub-centre, infield and CHC and available facilities in Varanasi district

S.No.	Staff	PHC (Block)	PHC (Add.)	Sub centre	Inplace/infield	CHC
1.	DMO incharge	–	–	–	1	–
2.	DVBDC consultant	–	–	–	–	–
3.	Block PHC MO /CHC MO	8	–	–	–	6
4.	Specialist	–	–	–	–	5
5.	PHC-MO	–	14	–	–	–
6.	Pharmacist	1	1	–	–	2
7.	Lab. Tech. (Regular)	8	–	–	–	1
8.	Lab. Tech. (Contract)	–	8	–	–	1
9.	Ward Boy	1	1	–	–	4
10.	Compounder	1	–		–	1
11.	Eye specialist	1	–	–	–	1
12.	Basic extension education	–	–	–	1	1
13.	Health visitors	–	–	–	4	1
14.	Health inspectors	–	–	–	3	1
15.	Health supervisors (Male)	–	–	–	8	1
16.	Health supervisors (Female)	–	–	–	42	–
17.	MPW (Male)	–	–	–	114	–
18.	MPW (Female)	–	–	–	332	–
19.	ASHA	–	–	–	1891	–
20.	ANM	2	1	1	–	3
21.	Malaria Technical Supervisor (MTS)	–	–	–	–	–
22.	Trained Dai	–	1	–	–	1
23.	Part Time Dai	–	1	–	–	1
24.	Sweeper	1	1	–	–	2
25.	Hospital assistant	–	–	–	1	–
26.	Lower clerk	1	–	–	–	–
27.	Computer	–	–	–	–	3
28.	Clerk	–	–	–	–	4

Source: CMO Office, Varanasi, 2010

have unsatisfactory condition of doctor's residence. It needs immediate enhancement of health facilities to rural population.

The regular supply of water and electricity, provision of telephone and vehicles are some of the essential requirement at each PHC for ensuring quick and better health care services, to the rural population. The availability of these infrastructural facilities portrays very poor and pathetic situation in the district. More or less all PHCs are supposed to refer the complicated cases to urban areas. The telephone facility at PHCs is very important but it is rarely available at each PHC of the study-area. It has been found that only 3 PHCs (Cholapur, Harhua and Kashi Vidyapith)

Table 4.6 Block-wise number of PHCs having specific infrastructure facilities

Blocks	Own building	Panchayat bhawan	Rental building	Pucca (Cemented) building	Doctor/resident	Tap water supply	Well	Hand pump	Electricity	Telephone	Vehicles
Baragaon	5	–	1	6	3	2	–	4	6	–	–
Pindra	2	–	–	2	1	1	–	1	2	–	–
Cholapur	3	1	–	4	1	2		2	4	–	1
Chirai Gaon	3	1	1	5	2	2	–	3	5	1	
Harhua	2	–	–	2	1	1	–	1	2	–	1
Sewapuri	2	–	1	3	1	1	–	2	3	1	1
Araziline	5	1	1	7	3	3	–	5	7	1	–
Kashi Vidhyapith	3	–	–	3	1	1	–	1	3	1	1
Total	15	3	4	32	13	13	–	19	32	3	4

Source: Based on Personal Survey, 2009

in rural areas of Varanasi district have telephone facilities. To make out-door programme of the PHCs a success, availability of vehicles in running is an essential requirement. Only four PHCs of the district have provision of vehicle to render health care facilities to the surrounding population (Table 4.7). Besides, provision of immunization to children and pregnant women is one of the important functions of PHCs. Therefore, for proper storage of essential vaccines at specific temperature refrigerator/freezers are storage also necessary. These are also used for storing blood for donation to serious patients.

It is found that only few PHCs have refrigeration's and freezers. To make these refrigerators operative, electric supply must be regular. This is quite good that each PHC have proper electric facilities. The continuous supply of water is another critical input. All PHCs posses in adequate water supply through taps. Only 13 PHCs have hand pump supply (Table 4.7). Unless and until these essential facilities are made available at the PHCs it is not possible to provide desirable services to the rural population living in influence zone of the PHCs.

Table 4.8 depicts the availability of specific services at PHCs. In order to fulfill the goal of National Population Policy, 2002 to provide 90 % institutional deliveries, availability of labor room is must at each PHCs. Similarly for providing reproductive and child health services the availability of pathological laboratory at every PHCs to conduct different types of tests are also essential. On an average only 56.3 %, 45.7 % and 35.1 % posses labor room sterilization and pathological testing facilities, respectively. The survey revealed that of the PHCs located in Baragaon, Pindra, Cholapur and Harhua blocks do not have labor room facility. In regards to sterilization facility PHCs at Baragaon and Sewapuri blocks PHCs have poor facility. It has been observed that PHCs at Kashi Vidyapith (66.6 %), Pindra (50.0 %) and Cholapur (50.0 %) blocks have high percentage of sterilization facilities. The district average is calculated as 45.7 % which is quite less than the desirable. It is further noted that PHCs at Pindra, Cholpaur and Harhua development blocks have provision of distributing Oral Pills and ORS packets along with availability of vaccines including BCG, OPV, DPT and measles injection.

Without availability of competent doctors, proper health care facilities could not be rendered to the neighboring rural population. During the survey it has been observed that only 86.45 % of the total PHCs of the district have doctors. The remaining PHCs are running without permanent doctors. It is heartening to note that PHCs at Chiraigaon, Harhua and Sewapuri development blocks lack in availability of doctors. The analysis of spatial distribution of health care facilities in rural areas of Varanasi district reflects that there are found inadequacy of infrastructural facilities for proper delivery of health care system to the rural population. This is the condition of PHCs in the study area, condition of sub-centre in Varanasi district in

Table 4.7 Block-wise percentage of PHC having specific infrastructural facilities

Blocks	Own building	Panchayat bhawan	Rental building	Pucca (Cemented) building	Doctor/resident	Tap water supply	Hand pump	Electricity	Telephone	Vehicles
Baragaon	83.3	–	16.6	100	50	33.3	66.6	100	–	–
Pindra	100	–	–	100	50	50	50	100	–	–
Cholapur	75	25	–	100	25	50	50	100	25	25
Chirai Gaon	60	20	20	100	40	40	60	100	–	–
Harhua	100	–	–	100	50	50	50	100	50	50
Sewapuri	66.6	–	33.3	100	33.3	33.3	66.6	100	–	33.3
Araziline	71.4	14.2	14.2	100	42.8	28.5	71.5	100	–	–
Kashi Vidhyapith	100	–	–	100	33.3	66.6	33.3	100	33.3	33.3
Total	78.1	9.3	12.5	100	40.6	40.6	59.4	100	9.4	12.5

Source: Based on Personal Survey, 2009

Table 4.8 Block-wise percentage of PHCs having specify services

Blocks	Labor room	Sterilization	IUD kit	Nirodh	Oral pills	Labour kit	Lab	BCG	OPV	DPT	Measles	Iron & folic acid	ORS packet	Obstetric drug kit	Specks
Baragaon	50.0	33.3	100.0	100.0	100.0	33.3	16.0	100.0	100.0	100.0	100.0	100.0	100.0	–	16.6
Pindra	50.0	50.0	100.0	100.0	100.0	100.0	50.0	100.0	100.0	100.0	100.0	100.0	100.0	–	50.0
Cholapur	50.0	50.0	100.0	100.0	100.0	100.0	50.0	100.0	100.0	100.0	100.0	100.0	100.0	25.0	25.0
Chirai Gaon	60.0	40.0	100.0	100.0	100.0	100.0	20.0	100.0	100.0	100.0	100.0	100.0	100.0	–	20.0
Harhua	50.0	50.0	100.0	100.0	100.0	100.0	50.0	100.0	100.0	100.0	100.0	100.0	100.0	50.0	50.0
Sewapuri	66.6	33.3	100.0	100.0	100.0	100.0	33.3	100.0	100.0	100.0	100.0	100.0	100.0	–	33.3
Araziline	57.1	42,8	100.0	100.0	100.0	85.7	28.3	100.0	100.0	100.0	100.0	100.0	100.0	14.2	14.2
Kashi Vidhyapith	66.6	66.6	100.0	100.0	100.0	100.0	33.3	100.0	100.0	100.0	100.0	100.0	100.0	33.3	33.3
Total	56.3	45.7	100.0	100.0	100.0	96.12	35.1	100.0	100.0	100.0	100.0	100.0	100.0	15.3	30.0

Source: Based on Personal Survey, 2009

also very pathetic than PHCs. For this purpose there is need of improving infrastructural facilities at all the PHCs of Varanasi district.

4.5 Proximity/Buffer Area Analysis of PHCs/New PHCs/CHCs

Proximity is the most basic GIS questions, such as:

- How close is this village to a hospital?
- Do any roads pass within 1000 m of a hospital?
- What is the distance between two hospitals?
- What is the nearest or farthest feature from any point of interest?

Buffer analysis is usually used to define protected zones around features or to show areas of influence. Buffers are sometimes used to clip data to a given study area, or to exclude features within a critical distance of something from further consideration in an analysis.

Buffer and Multiple Ring Buffer generate area features at a specified distance (or several specified distances) around the input features (point, line and polygon). Proximity analysis is used widely for many situations – e.g. to understand the association between transportation facilities in the study area to existing health care facilities This technique also provides answers to questions such as "how many persons live within a 10 km. radius from a particular primary health centre's (PHCs) or community health centres (CHCs) or from the other governmental hospitals etc. (Rai et al. 2011).

One way of defining accessibility to health centres is by knowing how far villagers live from their nearest centres. In this case, distance to provider is the main tool for measuring a Centre's accessibility. Based on the planning, standards of the Ministry of Municipal and Rural Affairs, every health centre should cover a catchments area extending 2 km radius wide. Accordingly, villagers residing within 2 km distance of a health centre are entitled to receiving services from this centre. ARC GIS-9.3 is used in the present study to define how much village fall within the catchments area of government health centres and which can be labeled as the served parts of study area. Three Buffer zone of 0–1000 m, 1000–2000 m and 2000–3000 m have been created around PHCs, New PHCs and CHCs of each development block of the study area to calculate the number of villages (to calculate the catchments are of the PHCs, New PHCs and CHCs) comes under in theses buffer zone of health centres (Fig. 4.8). Those villages who comes under these buffer zones are here called as safe village in terms of having health care facilities and those villages who did not fall under the buffer zone of PHCs, New PHCs and CHCs are called as unsafe villages. During the survey it is found the villagers who come under the safe villages travel at least 6 km to get medical treatment in government PHCs and CHCs and villagers in the unsafe villages can't easily approach at PHCs, New PHCs and

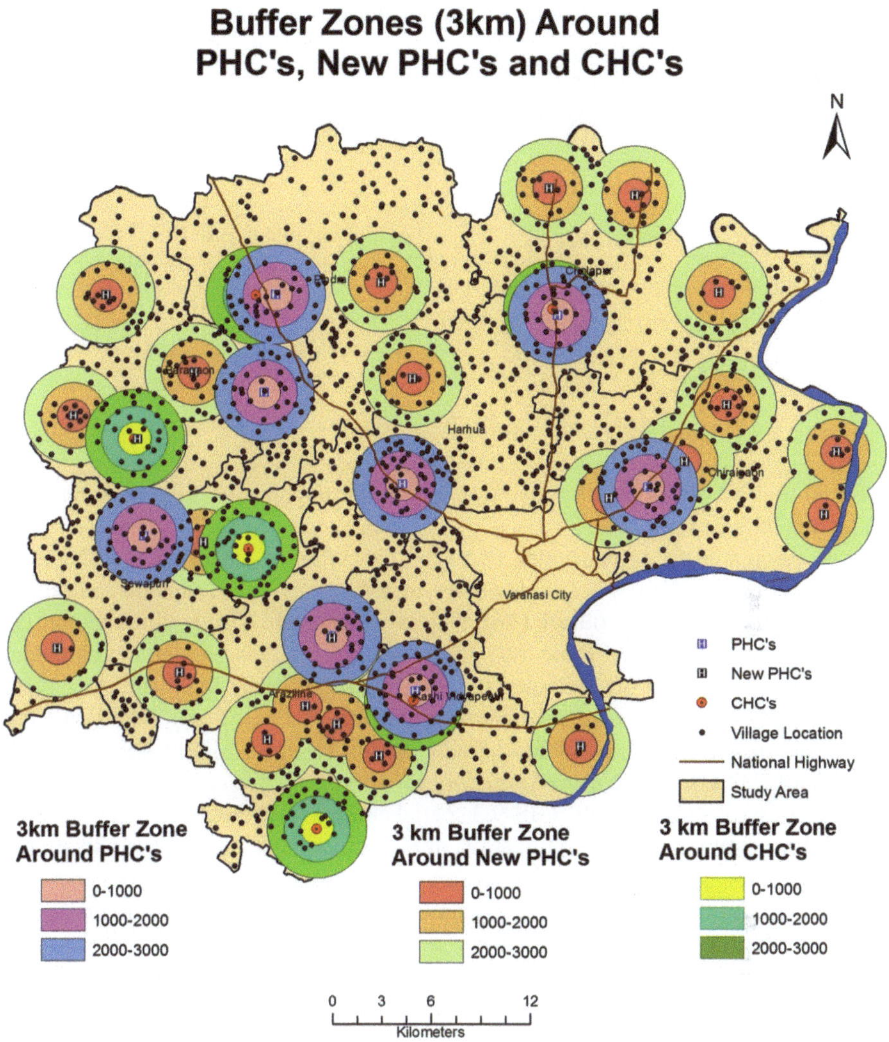

Fig. 4.8 Buffer zone (3 km.) around government health care units

CHCs to get any kind of medical treatments because of distance constraint. Each development blocks have many sub-centres but these sub-centres do not provide any good medical as well as infrastructure facilities to secure and help the rural people in any type of diseases or in any health related problems, so there is no mean to consider theses sub-centres for catchments area analysis. Rural people living in these villages always approach to PHCs or Newer PHCs or in CHCs for better medical treatment and facilities. By applying the proximity analysis of health centres, it

Table 4.9 Block-wise percentage of safe villages on the basis of buffer distance to PHCs, new PHCs and CHCs

Development block	Total village	Safe village	% of safe villages
Baragaon	185	136	73.51
Pindra	237	92	38.82
Cholapur	191	98	51.31
Chiraigaon	196	109	55.61
Harhua	212	89	41.98
Sewapuri	188	104	55.32
Araziline	257	168	65.37
Kashi Vidhyapith	146	45	30.82

Source: Self computed

is found that maximum rural population are totally depended on the existing government health centres.

It is calculated that, maximum 73.51 % (136 villages out of 185) villages in Baraagon development block are comes under safe village whereas minimum 30.82 % (only 45 villages out of 146) of villages in Kashi Vidyapith development block come in safe zones (Table 4.9). Only 52.17 % villages of each development block are comes under safe villages zone and they can easily approach to government health Centres for medical treatment and rest of the villages who don't fall under this buffer zone, considered here as unsafe villages. Figure 4.9 shows the location of safe villages in each development block.

The study selected the ARC GIS-9.3 spatial analyst to calculate distances around the health centres. For this application, the distance function selected to model proximity to the health centres computes an output raster dataset in which the output value at each location is potentially a function of all cells in the input raster data sets (McCoy and Jonhston 2001). The raster output of this function contains the measured distance from every cell to its nearest source. The distance function is potentially useful for such applications as finding the nearest hospital. It is also clear from this figure that existing health centres must serve a catchments area smaller than the standard size.

Health planners or concerned department can use this model in making decisions about where to build a new health centre in the study area. For example, areas beyond the 2Km accessibility zones can be used as a reference guide to enlist candidate locations for additional health centres in Varanasi district. The candidate locations can be presented to both regional and national health authorities for their consideration in building new health centres in selected parts of the study area (Rai et al. 2011).

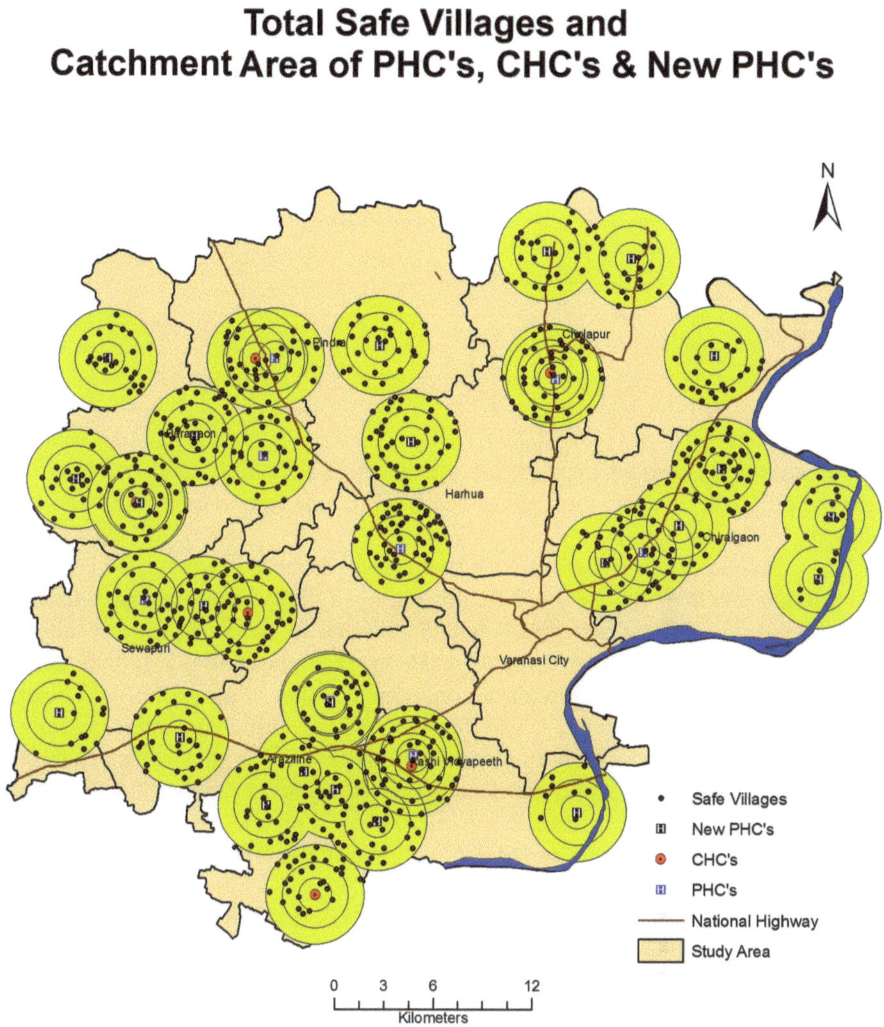

Fig. 4.9 Total safe villages comes in the buffer zones of government health facilities units in the development blocks of the Varanasi district

4.6 Shortest Road Estimation Through Network Analysis

Network analysis is used for identifying the most efficient routes or paths for allocation of services. This involves finding the shortest or least-cost manner in which to visit a location or a set of locations in a network. In all the solutions for static shortest path that unknown city, maybe the one which the probability of losing will be the least. Since the mobile users of transportation network want to find a path with

Table 4.10 Shortest distance between one government health care centre to another centre in each development blocks of Varanasi district

Development block	Stop count	Shortest distance between each health centre (km.)
Baragaon	5	34.80
Pindra	3	12.95
Cholapur	3	21.0
Chiraigaon	4	50.40
Harhua	2	17.62
Sewapuri	4	24.02
Araziline	6	35.14
Kashi Vidhyapith	2	12.14
Varanasi city	11	29.75

minimum time as cost, the optimal path is considered as finding a path between two specific points in transportation which needs minimum time to traverse (Dunn and Newton 1992).

The principle of equal access to health services for those in equal need is one of the guiding tenets of the National Health Service (NHS) in the Varanasi district. Nevertheless, health services are inevitably located in particular places and are therefore more accessible to nearby residents than those living farther away. Variations in proximity are, obviously, only one element of accessibility to health services (Ricketts and Savitz, 1994), but the physical difficulties of overcoming distance tend to be particularly important in rural area. Poor physical accessibility reduces the use of services and may lead to poorer health outcomes (Carr-Hill et al. 1997; Deaville 2001; Jones and Bentham 1997; Joseph and Phillips 1984). Low utilization of primary healthcare services is of particular concern Digital road network data of the study area are extracted from IRS-1C LISS III satellite data in conjunction with high resolution IKONOS satellite data. From the perspective of accessibility modeling, it is important that such details identify different classes of road (e.g., district road, national highway etc.) and ideally distinguish between sections in urban or rural areas. These road network data are very important to calculate the shortest distance between each hospital to the villages. Shortest distance between each Government health care units in each development blocks are calculated, which is based on Network analysis. In this table, Baragaon development block has shown 5 stop count (first health care unit to the fifth health care unit), and shortest distance between first stop to the last stop is 34.80 km. whereas in Kashi Vidhyapith block shown only 2 stop count to reach on health care units to another and shortest distance between both health care units is 12.14 km.

Details of measured shortest distance between government health care units in each development blocks are shown in the Table 4.10.

References

Aday, L. A., & Andersen, R. (1974). A framework for the study of access to medical care. *Health Service Research, 9*, 208–220.

Car-Hill, R., Place, M., & Posnett, J. (1997). Access and the utilization of healthcare services. In B. Ferguson et al. (Eds.), *Concentration and choice in health care*. London: Financial Times Healthcare.

Deaville, J. A. (2001). *The nature of rural general practice in the U.K.: Preliminary research*. Newtown: Institute of Rural Health and the General Practitioners Committee of the BMA.

Dunn, C. E., & Newton, D. (1992). Optimal routes in GIS and emergency planning applications. *Area, 24*(3), 259–267.

Jones, A. P., & Bentham, G. (1997). Health service and accessibility and details from asthma in 401 local authority districts in England and wales, 1988–1992. *Thorax, 52*, 218–222.

Joseph, A. E., & Phillips, D. R. (1984). *Accessibility and utilization: Geograpical perspectives on health care delivery*. New York: Harper and Row.

McCoy, J., & Johnston, K. (2001). *Using ArcGIS spatial analyst*. Redlands: Environmental Systems Research Institute, Inc.

Rai, P. K., Nathawat, M. S., & Singh, S. B. (2011). Role of GIS in analyzing the distribution of health care services. *National Geographical Society of India, 57*(3), 15–24. ISSN: 0027–9374.

Ricketts, T. C., & Savitz, L. A. (1994). Access to health services. In T. C. Ricketts, L. A. Savitz, W. M. Gesler, & D. N. Osborne (Eds.), *Geographic methods for health services research Lanham*. Lanham: University Press of America, Inc.

Chapter 5
GIS in Vector Born Disease Mapping

Abstract The representation and analysis of maps of vector born disease (VBD) and other related data is an important tool in the analysis and representation of local and regional variation in public health care system. GIS plays a variety of roles in the planning and management of the dynamic and complex healthcare system and disease mapping. Important vector born diseases like malaria, dengue fever, kala-azar etc. are discussed in this chapter. Spatial disease models study and predict the movements of people, information, and goods from one area to the other area. By accurately modeling these movements through GIS techniques, it is effortlessly to identify areas most at risk for disease transmission and thus target intervention efforts. Development block-wise report of VBD cases are mapped to recognize clusters necessitating intense attention for the control of disease. Location of dengue and kala-azar cases are identified through GPS. Important favorable indicators i.e. stream, ponds/water tanks, nalas, sewage zone, overhead tanks and slum areas in the Varanasi city also are very helpful malaria breeding sources and these indicators are extracted from remote sensing satellite data for the analysis. Outcomes of the present study recognized target variables that potentially favor mosquito breeding locations in the survey areas.

Keywords Vector Born Disease (VBD) • GIS • Remote sensing • Satellite data • GPS • Dengue • Malaria • Kala-azar • PHC • CHC

5.1 Vector Born Disease

"Vector-borne disease (VBD)" is the word mainly used to designate an illness due to an infectious microbe that is transmitted to any people through blood-sucking arthropods. VBD caused by a pathogen transmitted by an arthropod vector. Emerging and resurging vector-borne diseases (VBDs) cause significant morbidity and mortality, specifically in the developing world (42) (Eisen and Eisen 2011). The arthropods that most usually serve as vectors that comprise blood sucking insects such as mosquitoes, fleas etc. and blood sucking arachnids such as mites and ticks. A vector is an organism that does not cause disease itself but which transmits infection by conveying pathogens from one host to another. Vectors are facilitators of

P.K. Rai, M.S. Nathawat, *Geoinformatics in Health Facility Analysis*,
DOI 10.1007/978-3-319-44624-0_5

many perilous disease-causing organisms, the deterrence and treatment of which cannot be effective for long without addressing the vector directly. Vector densities are normally higher in rural than urban areas due to favorable habitats (Nipada 2005; Rai et al. 2011). The pervasiveness of geographical distribution of VBD has been progressively public health important and challenging problems in more than 100 tropical countries and it has been affecting more than 50 % of the population in the world, mainly, it is also a very main problem in India (Palaniyandi et al. 2014). The application of conformist and traditional method of vector control in the field has substantial effect, however the problem is not manageable and it has been increased progressively.

5.2 Disease Mapping

The representation and analysis of maps of disease and other related data is an important tool in the analysis and representation of local and regional variation in public health care system. In the case of disease spread, individuals near or incontact to a contagious person or a contaminated environmental setting are deemed more susceptible to certain types of illnesses (Rai et al. 2012). Cartographic mapping techniques play a significant role in the representation, display and study of the geographical data. Cartographic techniques have the following steps: (i) Geographical Feature Classification, (ii) Scale Determination, (iii) Symbol Categorization, and (iv) Graphic Primitives (Fig. 5.1) (Lai et al. 2009). The elementary working units of disease data include point (e.g. patient house locations), line (e.g. transmission route), and area (e.g., disease rate). Recently, GIS has emerged as an important element of various projects in public health and epidemiology (Foley 2002). GIS plays a variety of roles in the planning and management of the dynamic and complex healthcare system and disease mapping. Although still at an early stage of integration into public healthcare planning, GIS has shown its capability to answer a diverse range of questions relating to the key goals of efficiency,

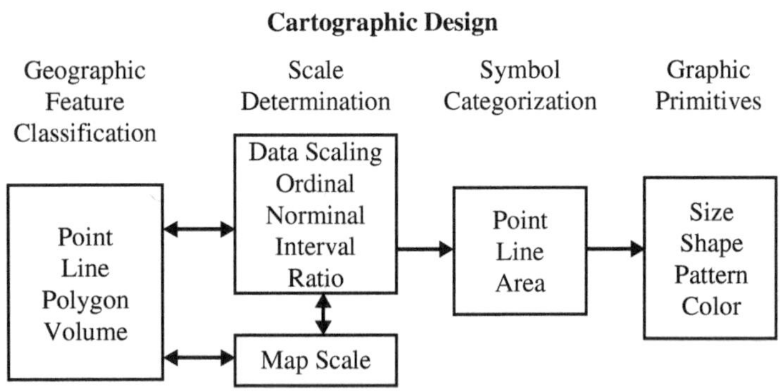

Fig. 5.1 A framework of cartographic design (Source: Lai et al. 2009)

effectiveness, and equity of the provision of public health services (Kleinschmidt et al. 2000; Boscoe et al. 2004). Conclusively, GIS will play a significant part in the reorganization of public health and disease planning in the twenty-first century (Kamel Boulos et al. 2001), particularly in response to sweeping changes taking place in the handling of health information.

Epidemiologists have usually used maps when studying associations between location, environment, and disease (Rai et al. 2011). Technological advances over the last decades with significance to VBDs include the emergence of molecular techniques for vector species identification and pathogen detection and identification, and a rapid development in hardware and software systems to support data collection, management, analysis and data integration with other data formats (Eisen and Eisen 2011).

GIS has been used in the surveillance and monitoring of vector-borne diseases, water-borne diseases, in environmental health, analysis of disease policy and planning, health situation in an area, generation and analysis of research hypotheses, identification of high risk health groups, planning and programming of activities, and monitoring and evaluation of interventions (Gatrell and Loytonen 1998; Gatrell 2002; Rai et al. 2011). GIS-based disease mapping has been used for risk assessment at national, regional, town and village level. GIS-based statistical models used to estimate vector presence or abundance, or VBD case presence or incidence, within a particular geographical area (Eisen and Eisen 2011).

Mapping disease distribution, spatial analysis and spatial modeling of VBD are progressively essential now days at both regional and global level, and the results obtained using geospatial technologies i.e. remote sensing, GIS and GPS has potentially been important role mostly in the designing and implementation of active disease control programmes, has already been used to representative and implementing the programme for malaria (Palaniyandi et al. 2014; Palaniyandi and Nagarathinam 1997; Mushinzimana et al. 1997).

Such mapping is measured crucial for analyzing past as well as present disease trends. Many development agencies and government institutions are exploring Health GIS in India. However, the sheer size of our country, varied life styles, climatic zones and environmental conditions make it all the more important for India to have a health GIS.

5.3 Important Vector-Borne Diseases

5.3.1 Malaria

Malaria (marsh fever, periodic fever) is a parasitic disease that comprises infection of the red blood cells (RBCs). Park (1972) expresses malaria as a communicable disease caused by sporozoan parasites of the genus *Plasmodium SP* & transmitted to man by certain species of infected female *Anopheles mosquitoes*. Malaria exists

Plate 5.1 Mosquito breeding sources

in every tropical and subtropical area across the globe, sometimes making seasonal excursions into temperate regions as well (Rai et al. 2011, 2013).

The protozoan parasites that cause it have more complex genomes, metabolisms and life cycles than almost any other vector-borne threat. This difficulty makes them a tough target for interferences such as drugs and vaccines because the parasite's shape-shifting ways allow it to evade chemical and immunological defenses (Rai et al. 2011, 2012). Malaria mosquito breeding sources in the study area are shown in Plate 5.1.

5.3.1.1 Factors Responsible for Occurrence of Malaria Disease

Related Climatic Factors

The climatic conditions have a profound consequence on the life cycle and longevity of vector mosquito and also on the growth of the malaria parasite (Craig et al. 1999). Climate has vital role within thresholds for mosquito and parasite survival. The climate change (temperature, rainfall and humidity, saturation deficit (SD) are significant factors of vector breeding and survival (Bhattacharya et al. 2005, 2006; Palaniyandi 2013). The essential climatic factors are:

- Temperature and Relative Humidity (RH)

The optimal temperature for development of malaria parasite is between 25 and 30 °C, and average RH at least 60 %. These temperature and RH conditions increase the permanency of mosquitoes & thus help malaria transmission.

- Rainfall or Precipitation

Increased rainfall particularly monsoon days result in mosquitogenic situations. Pre-monsoon rains, which maintain temperatures between 25–30 °C and relative humidity around 80 % for 25–30 °C longer duration, lead to increase vector density and durability to initiate malaria transmission, if parasite load occurs in the community.

Natural Calamities

The natural calamities/disasters like flood, drought and earthquake, usually because increase in mosquitogenic conditions result in outbreak of malaria epidemic.

Vulnerability

Some of the significant points of vulnerability of an area are as follows:

- In urban slums area, growth in mosquitogenic conditions due to increased houses and settlement, water storage, poor drainage and sanitation facility;
- Change in bionomics and behavior of vector/s with special reference to change in resting and feeding habits, and response to insecticides;
- Deforestation in the area;
- Increase in breeding and density of vector and Increase in the degree of man-mosquito contact.

Mosquito and Parasite Factors

- Growing trend of fever cases and SPR (Slide positivity rate);
- Growing in proportion of gametocytes in the community;
- Resistance in parasite to drugs;
- Growth in malaria mortality; and
- Increase in consumption of anti-malaria's

Related Operational Factors

Insufficient basic health care services may contribute heavily to outbreak of malaria epidemic in the area, mainly the following points:

- Staff (medical and para-medical) position;
- Poor surveillance services;
- Lack of appropriate transportation facilities;
- Insufficiency in material, machines and related medical equipment; and
- Insufficient advance training and lack of inspiration of work.

5.3.2 Dengue Fever

Dengue virus in many ways appears to be the obverse of malaria. Malaria transmission arises most often in rural areas while Dengue is a city disease. Although the *Anopheles* vectors of malaria bite mostly at night, the *Aedes* vectors of dengue bite mainly in the daytime. While an initial malaria infection generally produces the most severe symptoms, a second infection of dengue can be much more dangerous than the first, when it involves a different serotype of the virus (Rai et al. 2011). Dengue fever can be painful (hence its nickname of "break bone fever") and debilitating but is generally not life threatening when first acquired. However, severe manifestations arise in areas where more than one of the four main strains of the virus coexists. Becoming exposed to a second, different strain of the virus can provoke a severe immunological reaction called Dengue Hemorrhagic Fever (DHF), which can carry a significant risk of death, especially in children and younger adults. Presently, about 5 % of the hundreds of thousands of people who acquire DHF die, although prompt and effective medical care can greatly reduce this case fatality rate (Rai et al. 2011).

5.3.3 Filariasis

Filariasis generally doesn't kill but it can cause considerable disability. Several forms of this mosquito-borne infection are caused by nematode worms that invade the lymphatic system causing swelling and tissue buildup in various parts of the body, but particularly affecting the legs (Rai et al. 2011). In its most serious manifestation, this disease causes grotesque distortion of appendages known as "elephantiasis." Major surgery and extensive tissue removal provides the only cure for infections that reach this level of severity. The more severe symptoms require many years of repeated infection to develop, thus would not present a significant problem to workers in the short-term. However, on-the job exposure can cause many years of declining health and suffering (Rai et al. 2011).

5.3.4 Kala-Azar

Kala azar (*Visceral Leishmaniasis*) is a deadly disease caused by the parasitic protozoa *Leishmania donovani* and transmitted to humans by the bite of infected female sand fly, *Phlebotomus argentipes*. The amastigote form of the parasite invades the Reticulo Endothelial system of humans. It lowers immunity, causes persistent fever, anemia, liver and spleen enlargement, and if left untreated, it kills. The vector thrives in cracks and crevices of mud plastered houses, poor housing conditions, heaps of cow dung, in rat burrows in bushes and vegetation's around the houses (Rai et al. 2011).

5.4 Data Required for GIS Based Vector Born Disease (VBD) Mapping

A GIS encompasses three types of information files: geographic map, attribute, and data-point files. Different spatial and non-spatial sets of data are to be used in this study. District boundary map as well as the development blocks and village boundaries maps are used which are collected from 'Land Survey Revenue Office', Varanasi; U.P. Disease incidence report from different years is used for this study. The data is collected from District Malaria Office of Varanasi, U.P., India and the population data of 2001 is used for this study. Physical altitude map is produced from the SOI topographical sheet (1:50,000 scale) of Varanasi district (Rai et al. 2011).

In general, modeling comprises the integration of GIS with statistical and health science approaches. Spatial disease models study and predict the movements of people, information, and goods from one area to the other area. By accurately modeling these movements through GIS techniques, it is effortlessly to identify areas most at risk for disease transmission and thus target intervention efforts. Spatial disease models examine and predict the spread of phenomena over space and time and have been extensively used in understanding spatial diffusion of diseases. By integrating a temporal dimension, these models can calculate how diseases spread, spatially and temporally from infected to susceptible people in an area. Spatial variation in health related data is well known, and its study is a vital aspect of epidemiology (Rai et al. 2011).

5.5 Methodology Adopted for Vector Born Disease (VBD) Mapping

All the maps are properly rectified to real world co-ordinate system with respect to well-known reference point (ground control point), based on UTM Zone 44 projection system and WGS-1984 datum. Spatial data are digitized using ARC GIS -9.3 software. All the coverage's are editing to eliminate digitization errors such as overshoot, undershoot, dangles, and labels for each polygon features. Database is produced based on disease incidence report and related with the GIS vector layer. Digital maps are created on GIS Platforms and suitable cartographic techniques. For GIS platform geo-referenced digital map of villages/tehsils/districts are used. A three level database is created i.e. district wise, tehsil wise and village-wise. Attribute data such as village wise population, medical facilities, primary health centers, etc. and data on VBD for 5 years – from 2005 to 2009, are attached to the village maps and are used for the analysis for decision support in preparation of control strategies (Rai et al. 2011). The significant types of vector born disease (VBD) highlighted in this study are malaria, filarial, kala-azar, jondis, dengue. Data

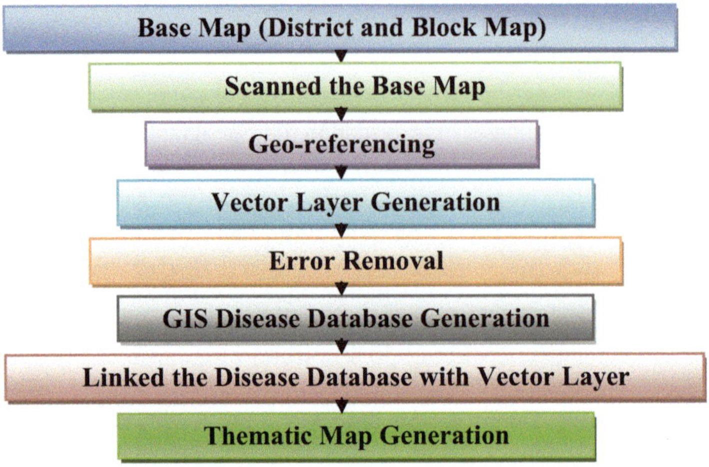

Fig. 5.2 Flowchart of the methodology for disease mapping

analysis has also shown that many case of VBD has been identified in the study area from 2005 to 2009 (Fig. 5.2).

5.6 Status of Vector Born Diseases (VBD) in the Varanasi District

Application of GIS in epidemiology through envisaging and studying geographic distribution of diseases through time, thus revealing spatio-temporal trends and patterns that would be more tough or obscure to discover in other way (Nipada 2005). It performs a spatial statistical task such as overlaying of different layers of information to determine dependencies and relationship between outbreaks and environmental factors. Climate plays a key role in disease occurrence. Higher temperature and humidity is essential for the malaria, dengue and kala-azar vector breeding. The annual maximum and minimum temperature ranges from 35–32 °C and 10–12 °C respectively and humidity ranges from 70 to 80 % in the study site and the annual average rainfall of the district is 1154 mm. These types of climate characteristics are responsible for the vector breeding (Napier 1926). The density of vector increases in summer and rainy season and in the winter season the density is less. Geo-spatial study delivers the information of land cover, soil, terrain, and demographic characteristics (Thakur 2000) of the area i.e. highly influenced the vector of the disease. The year wise distribution of the disease and annual number of human affected by vector born disease from the year 2005 to 2009–2010 are shows in Table 5.1.

Till November 2009–2010, a total of 2145 cases of filarial and 13 case of kala-azar case are identified in the area and out of which majority of Kala-azar case are

Table 5.1 Year wise report of Vector Born Disease (VBD) in Varanasi district

Year	Malaria	Filaria	Kala-azar	Dengue
2005	226	2212	–	09
2006	296	2236	13	06
2007	277	2474	27	38
2008	322	2327	22	15
2009–2010	617	2145	13	96

Source: DMO Office, Varanasi District

Table 5.2 Status of Kala-azar diseases, 2006–2010

CHC/PHC/	No. of Kala-azar Cases				No. of the Sub-centre and Village			
Urban Area	2006	2007	2008	2010	2005	2006	2007	2009–2010
Urban	–	02	–	05	–	Ramnagar	–	Orderly Bazar, Taktakpur
Harahua	–	–	06	02	–	–	Tewar-2	Lamahi-1, Soyepur, Tewar-1
Sewapuri	–	–	–	–	–	–	–	–
Chiraigaon	01		01	01	Chhitauni		Chhitauni	Kadipur-Bankat
Araziline	01	–	–	–	Mahgaon, Chandapur	–	–	–
Cholapur	–	01	01	–		Bhawanipur	Cholapur Proper	–
Kashi Vidyapith	11	24	14	05	Nathupur-Bhullanpur	Nathupur-Bhullanpur, Main Centre-Maheshpur, Harpalpur	Harpalpur, Rahimpur, Govindpur	Harpalpur-3, Khulashpur-2
Total	13	27	22	13	–	–	–	–

Source: District Malaria Office, Varanasi District and through GPS, 2010 Rai et al. (2011)

found in Harpalpur (03) and Khulashpur village of (02) villages of Kashi Vidyapith development block while two and one kala-azar patient is identified in Harhua and Chiraigaon block respectively. Five kala-azar patients are diagnosed in the urban area especially in Orderly Bazaar and Taktakput urban health centres. The kala-azar cases looked to be decreasing during year 2006 to 2009–2010 but malaria and dengue patients are viewing growing trend. Analyzing the disease incidence data, it is perceived that the maximum case of kala-azar are recorded in 2007 (27), 2008 (22), and it is gradually decreased in 2009–2010 (13).

Annual filarial case total ranged from 2212 in 2005 to 2145 in 2009, with a trend of little bit declining incidence. Maximum cases of filarial are found in the year

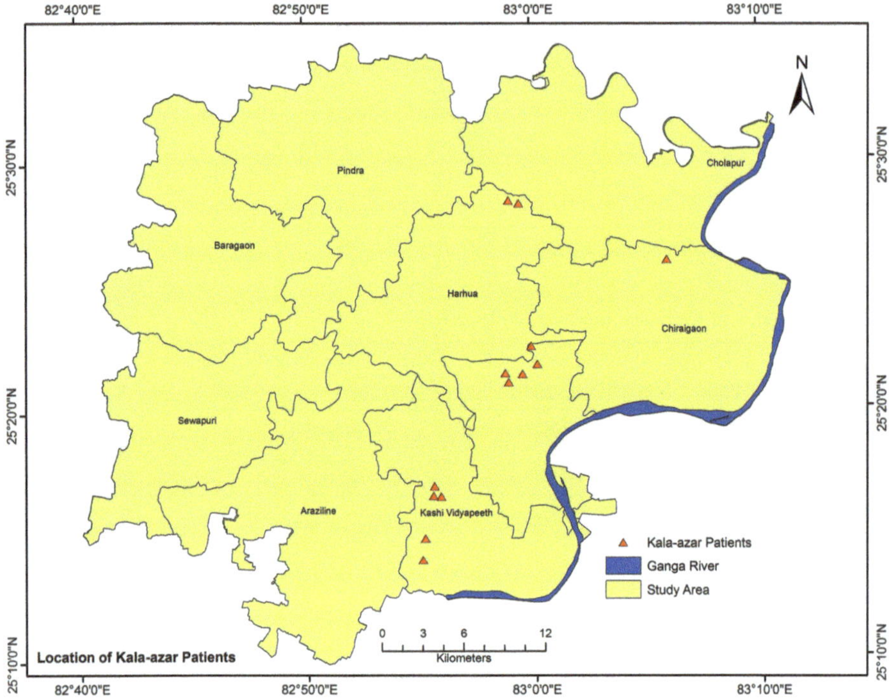

Fig. 5.3 GPS location of Kala-azar patients in Varanasi district (2009–2010)

2007, which is 2474. Situation analysis of kala-azar disease status PHC's/CHC's wise is given in the Table 5.2. Location of kala-azar patients is shown in the Fig. 5.3 (Rai et al. 2011).

Block-wise report of VBD cases are mapped to recognize clusters necessitating intense attention for the control of disease. Dengue viruses have been predictable in recent decades as one of the most dreadful diseases global. Dengue fever is a life intimidating disease if a person is being diseased more than once. Varanasi district recorded 96 suspected cases of dengue fever in 2009–2010 (Rai et al. 2011).

However the number of cases is not high compared with some tropical countries, preventive measures and surveillance strategies against the sporadic occurrences of dengue fever are still necessary, especially in a densely populated study area. Dengue disease data are collected and considered here only from important government hospitals in Varanasi city e.g. Sir Sunderlal hospital B.H.U., Pandit Deen Dayal hospital and Shiva Prasad Gupta hospital. Maximum cases of dengue are identified in urban area of Varanasi district where sanitation and ponds condition are not very much adequate. Three cases of dengue fever are identified in Kashi Vidyapith development block in which two cases came at Mishirpur CHC and one case is recorded in PHC of this block. Single case of Dengue fever is also recorded in Chiraigaon, Cholapur and Araziline PHC respectively. No. of dengue fever cases in Varanasi district is shown in the Fig. 5.4 (Rai et al. 2011).

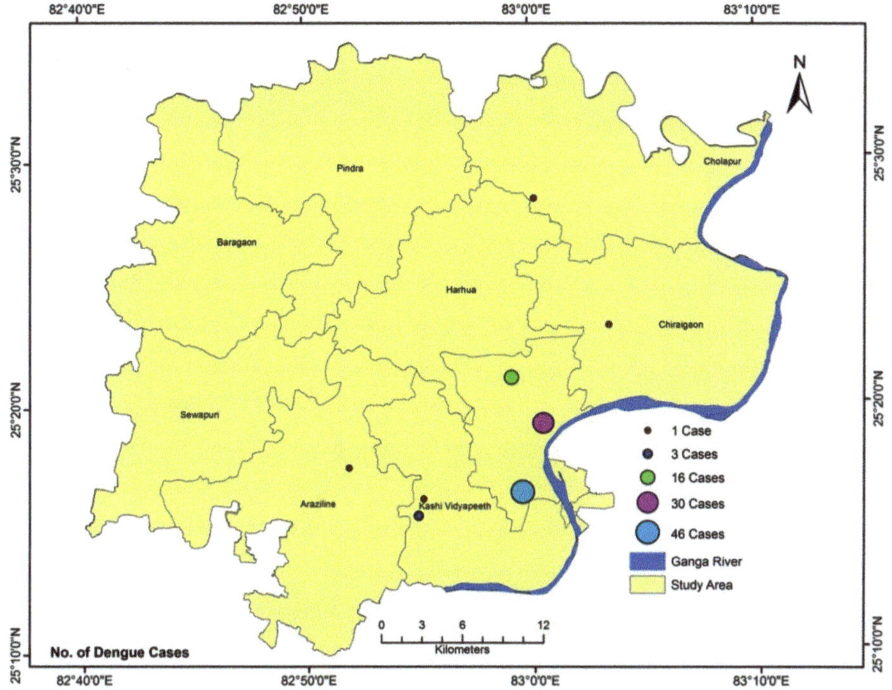

Fig. 5.4 PHC's and hospital-wise dengue cases in Varanasi district (Rai et al. 2011)

Table 5.3 Block-wise status of malaria patient in Varanasi district

	2005		2006		2007		2008		2009	
PHC/CHC	B/S Exa.	± ve	B/S Exa.	± ve	B/S Exa.	± ve	B/S Exa.	± ve	B/S Exa.	± ve
Baragaon	2306	61	3902	85	3959	105	2818	117	1645	126
Pindra	6067	58	7454	67	7118	21	7436	44	6422	37
Cholapur	4157	23	4912	20	3887	60	4293	72	3846	92
Chirai Gaon	4071	36	6193	18	5647	17	5234	23	4777	41
Harhua	3305	06	5992	14	6077	17	6927	11	4235	35
Sewapuri	2635	20	4315	46	2836	7	2572	5	1933	16
Araziline	3374	03	6553	7	7082	20	6001	11	6117	15
Kashi Vidhyapith	3432	04	4245	5	3439	9	2764	6	2962	67
Urban	1561	09	1525	34	1896	21	3167	33	3178	188
Total	31,807	226	45,091	296	41,941	277	41,257	322	35,115	617

Source: DMO Office, Varanasi District, 2009. (B/S Exam.-Blood Sample Examined)

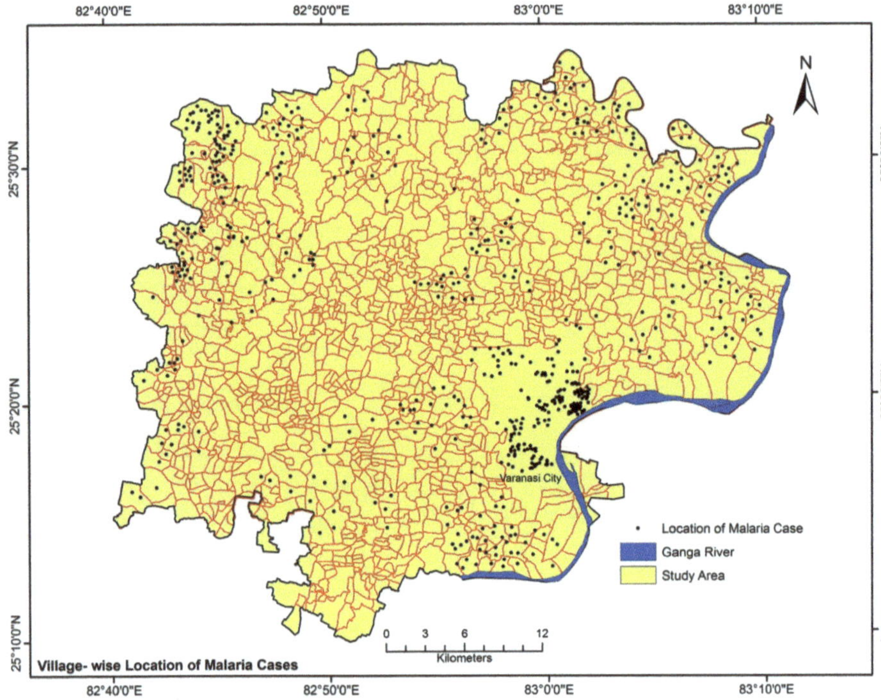

Fig. 5.5 Village-wise location of malaria cases collected through GPS

In Varanasi city, most of the houses are pucca (cemented) types and had air coolers. General sanitary conditions around most of the houses are found not very well. Water storage period in some areas is for 3–6 and >6 days. The key containers in study area are overhead tanks, underground tanks, ground level tanks and buckets. In the highest incidence areas, emptying/drying of water containers is mostly done on weekly basis. From the Table 5.3 it is shown that in year 2008 and 2009 maximum number of malaria positive cases has been identified, which is 322 and 617 respectively. Village-wise location of malaria cases are located through GPS and shown in the Fig. 5.5. In the year 2009–2010, maximum cases of malaria patients are identified in Baragaon (126) development block followed by Cholapur (92) and Kashi Vidyapith (67) whereas only 15 malaria cases are recorded in Araziline block. Status of malaria patient in year 2006–2009 is shown in the Fig. 5.6a–d. Important favorable indicators i.e. streams, ponds/water tanks, nalas, different sewage zone, overhead tanks and slum areas in the Varanasi city also are very helpful malaria breeding sources and are extracted from Remote Sensing satellite data. All these indicators are shown in the Fig. 5.7 (Rai et al. 2011).

In cases of concrete housing, the adjacent cattle barns are made of thatched roofs, which provided an appropriate place for the breeding of mosquitoes. The usage of open ground tanks for water storage afforded another major breeding site. The lack of consciousness of proper sanitation services and avoidance behavior also added to

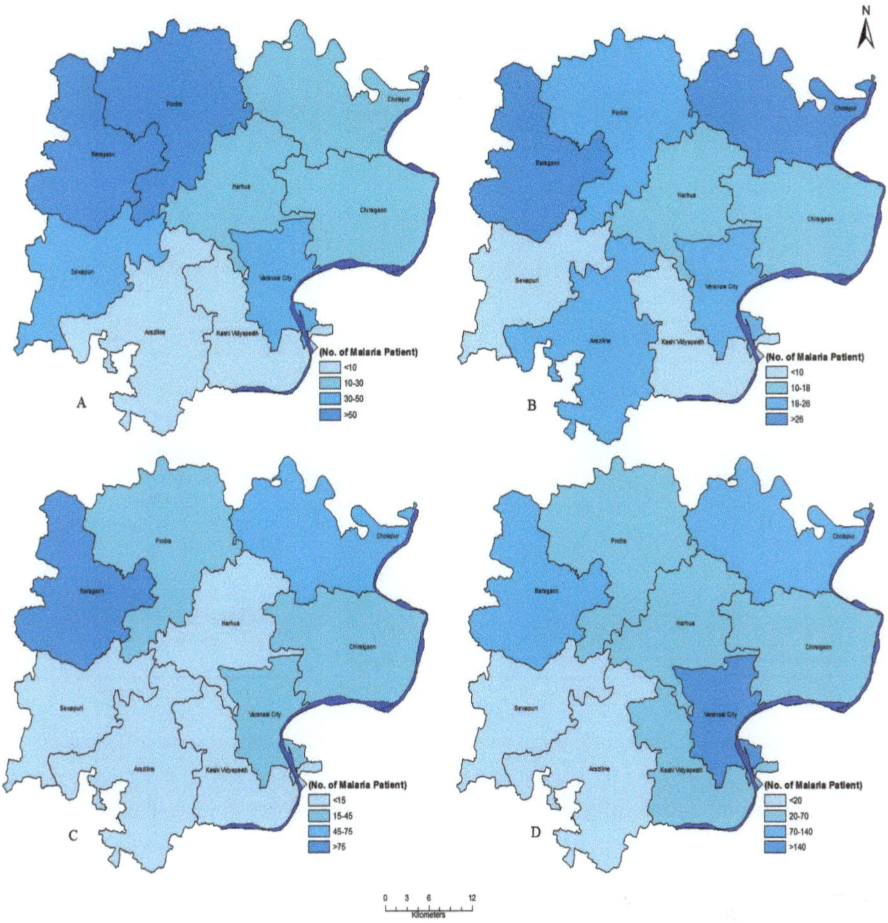

Fig. 5.6 (**a**) to (**d**) Status of malaria patients in Varanasi district in 2006 (**a**), 2007 (**b**), 2008 (**c**) and 2009 (**d**) respectively (Rai et al. 2011)

the context. In addition, it is well-known that people stayed outside their houses utmost of the time i.e., they slumbered outside their houses in the evenings or at night when mosquito activity is at its peak. These residents are mostly farmers who wore partial or improper clothing to facilitate the daytime heat. Mostly people reared cattle's and other livestock's within their premises which also led to constant and favorable conditions for mosquito proliferation. In the infected areas, the schools are located closer to farms or had water storage tanks nearby. This exercise controlled to the children being infected more than the adults.

Outcomes of the present study recognized target variables that potentially favor mosquito breeding locations in the survey areas. The results should help in rallying social behavior and communication programs that may be applied to reduce the breeding sites based on community involvement. Additionally, resource allocation

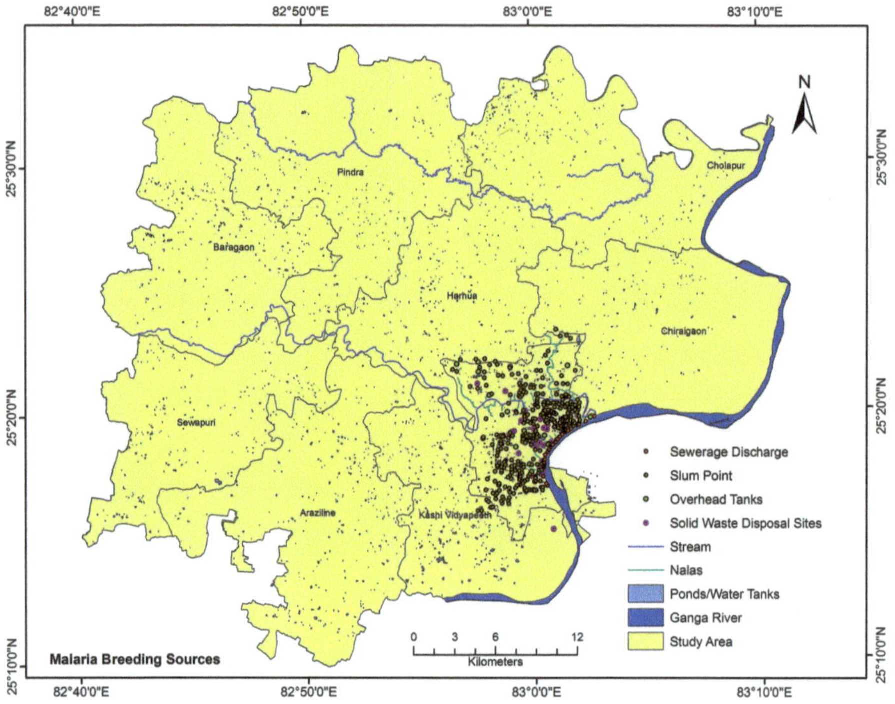

Fig. 5.7 Important indicators of malaria breeding source extracted from remote sensing data

can be recommended to infected areas identified through population-based-surveys, which can be conducted as a component of malaria control surveillance. Villages with primary health centers are mapped and a buffer zone is generated at 3 km and village boundaries are overlapped to show the accessibility for patients to closest PHC's (Rai et al. 2011).

References

Bhattacharya, S., Chakravarty, S., Maity, S., Dureja, V., & Gupta, K. K. (2005). Metal contents in the groundwater of Sahebgunj district, Jharkhand, India, with special reference to arsenic. *Chemosphere, 58*, 1203–1217.

Bhattacharya, S., Sharma, C., Dhiman, R. C., & Mitra, A. P. (2006). Climate change and malaria in India. *Current Science, 90*(3), 369–375.

Boscoe, F. P., Ward, M. H., & Reynolds, P. (2004). Current practices in spatial analysis of cancer data: Data characteristics and data sources for geographic studies of cancer. *International Journal of Health Geography, 97*, 14041–14043.

Craig, M. H., Snow, R. W., & Le Sueur, D. (1999). A climate-based distribution model of malaria transmission in sub-Saharan Africa. *Parasitology Today, 15*, 105–111.

Eisen, L., & Eisen, R. J. (2011). Using geographic information systems and decision support systems for the prediction, prevention, and control of vector-borne diseases. *Annual Review of Entomology, 56*, 41–61.

Foley, R. (2002). Assessing the applicability of GIS in a health and social care setting: Planning services for informal carers in East Sussex, England. *Social Science and Medicine, 55*, 79–96.

Gatrell, A. C. (2002). *Geographies of health: An introduction*. Oxford: Blackwell Publishers.

Gatrell, A., & Loytonen, M. (1998). *GIS and health*. London: Taylor & Francis.

Kamel Boulos, M. N., Roudsari A. V., & Carson E. R. (2001). Health geomatics: An enabling suite of technologies in health and healthcare (Methodolical Review). *Journal of Biomedical Information, 34*(3), 195–219. doi:10.1006/jbin.2001.1015; http://dx.doi.org/10.1006/jbin.2001.1015

Kleinschmidt, I., Bagayoko, M., Clarke, G. P. Y., Craig, M., & le Sueur, D. (2000). A spatial statistical approach to malaria mapping. *International Journal of Epidemiology, 29*, 355–361.

Lai, P. C., So, F. M., & Chan, K. W. (2009). *Spatial epidemiological approach in disease mapping and analysis* (pp. 1–169). Boca Raton: CRC press.

Mushinzimana, E. S., Munga, N., Minakawa, L. L., Feng, C. C., Bian, L., & Kitron, U. (1997). Landscape determinants and remote sensing of anopheline mosquito larval habitats in the western Kenya highlands. *Indian Malaria Journal, 5*(1), 13.

Napier, L. E. (1926). An epidemiological consideration of the transmission of Kala-azar in India. *Indian Medical Research Memoirs, 4*, 219–265.

Nipada, R. (2005). GIS and epidemiology. *Journal of Medical Association, Thai, 88*(11), 1735–1738.

Palaniyandi, M. (2013). GIS mapping of vector breeding habitats. *Geospatial World Weekly, (GIS e-news magazine), 9*(2):1–4.

Palaniyandi, M., & Nagarathinam, V. (1997). Land use/land cover mapping and change detection using space borne data. *Journal of the Indian Society of Remote Sensing, 25*(1), 27–33.

Palaniyandi, M., Anand, P. H., & Maniyosai, R. (2014). Spatial cognition: A geospatial analysis of vector borne disease transmission and the environment, using remote sensing and GIS. *International Journal of Mosquito Research, 1*(3), 39–54.

Rai, P. K., Nathawat, M. S., Mishra, A., Singh, S. B., & Onagh, M. (2011). Role of GIS and GPS in VBD Mapping: A Case Study. *Journal of GIS Trends. Academy Science Journals, North America, 2*(1), 20–27. ISSN:2146 0892.

Rai, P. K., Nathawat, M. S., & Onagh, M. (2012). Application of multiple linear regression model through GIS and remote sensing for malaria mapping in Varanasi District, INDIA. *Health Science Journal (International Journal of Nursing Research and Review), 6*(4), 731–749. ISSN 1108-7366E. Available at http://www.hsj.gr/volume6/issue4/6414

Rai, P. K., Nathawat, M. S., & Rai S. (2013). Using the information valuemethod in Geographic Information System (GIS) and remote sensing for malaria mapping. *A case study from India, Informatics in Primary Care (University of Surrey, U.K.), 21*(1), 43–52. ISSN 1475-9985. Available at http://dx.doi.org/10.14236/jhi.v21i1.38

Thakur, C. P. (2000). Socio-economics of visceral leishmaniasis in Bihar (India). *Transactions of The Royal Society of Tropical Medicine and Hygiene, 94*, 156–157.

Chapter 6
A Study of Malaria Susceptibility Mapping Using Statistical Methods with GIS

Abstract Developments in the area of geographical information systems (GISs) can offer new ways to symbolize epidemiological data spatially. In this chapter, three statistical methods i.e. Multiple Linear Regression, Information Value (Infoval) and Heuristic Method are used to develop malaria susceptibility index (MSI) and malaria susceptibility zonation (MSZ) through GIS and remote sensing. Village-wise malaria location data were collected from each primary health centre (PHC) and then the locations were identified using GPS. Malaria influencing data layers like rainfall, temperature, population density, distance to river, distance to roads, distance to health facilities, distance to ponds, NDVI, land use. are very well described in this study though GIs and remote sensing. These layers are used to produce the malaria-susceptibility model map. Comparison of statistical methods to select optimum model for MSZ by malaria density method (*Qs*) is also calculated. The malaria investigation is also completed using the information value, multiple linear regressions, and heuristic models, and the analysis outcomes are tested using the malaria locations for the study area. The verification method is achieved by Area under Curve (AUC).

Keywords GIS • Multiple linear regression • Information value (Infoval) and heuristic method • Remote sensing • Malaria Susceptibility Zonation (MSZ) • NDVI

6.1 Introduction

Malaria transmission depends on various features that all have an influence on the vectors, parasites, human hosts and the interactions among them. These factors may include, among others, meteorological and environmental conditions (Roper et al. 2000). The most apparent determinants are detected to be the meteorological and environmental parameters such as rainfall, temperature, humidity and vegetation type and cover (Connor et al. 1997; Craig et al. 1999). There are only a few examples of the application of epidemiological maps in malaria control and this may be explained by a lack of suitable, spatially defined data and also by a relatively incomplete understanding of how epidemiological variables relate to disease occurrence. Recent evidence suggests that clinical manifestations of infection are determined by

P.K. Rai, M.S. Nathawat, *Geoinformatics in Health Facility Analysis*,
DOI 10.1007/978-3-319-44624-0_6

the intensity of parasite exposure. Developments in the area of geographical information systems (GISs) can offer new ways to symbolize epidemiological data spatially (Omumbo et al. 1998).

The role of Geographical Information Systems to the understanding of requirements assessment and health care is a relatively new advance, but the geographical study of disease and health when dealing with such environmental substances as the tracing of different epidemics (Colledge et al. 1996). GIS software is being used to correlate the climatic attributes of the collection localities with the presence or absence of various mosquitos' parasites (Sweeney 1997). This technology has been in existence for a number of years, but it is only recently that it has been extensively accepted as a powerful tool to augment existing monitoring and evaluation approaches (Connor et al. 1997; Sudhakar et al. 2006). Vegetation is often associated with vector breeding, feeding, and resting locations. A number of vegetation indices have been used in remote sensing and Earth science disciplines. Vegetation has often been found associated with the vector's breeding, feeding and resting locations. A number of vegetation indices have been used in remote sensing and earth science disciplines. The most widely used index is the Normalized Difference Vegetation Index (NDVI). It is well-defined as the difference between red and near-infrared bands normalised by double the mean of these two bands. For green vegetation, the reflectance in the red band is low because of chlorophyll absorption, while the reflectance in the near-infrared band is high because of the spongy mesophyll leaf structure (Hoek et al. 2003).

Mathematical and statistical components embedded in GIS enable the testing of hypotheses and the assessment, explanation, and prediction of spatial and temporal tendencies (Kulldorff 1998; Lawson 2009). Statistical techniques model the relation between parasite exposure risk and environmental risk aspects via a multivariate linear regression model. These can also be used for prediction (Riedel et al. 2010, 2013; Rai and Nathawat 2013).

The main objective of malaria-susceptibility modelling is to observe the long-term parameters, and to define their relations to malaria occurrence in the studied area (Rai et al. 2013; Rai and Nathawat 2013). The primary objective has been set for assessment and appraisal of significant parameters for incidence of malaria disease in the study area and, then selection of the utmost determinant parameters. The final aim of the research is malaria susceptibility zonation in the study area. In this study, a number of modeling methods comparatively used for MSZ so that an optimum model for the study area can be selected and tested for its suitability (Rai et al. 2013; Rai and Nathawat 2013).

6.2 Data Collection and Methodology for Malaria Susceptibility Modeling

A number of thematic maps (referred to as data layers in GIS) were generated on specific parameters associated to the consequences of malaria, that is land use, normalized difference vegetation index (NDVI), distance to water bodies (such as ponds, rivers etc.), roads and health centres, rainfall and temperature data and projected population density in 2009. In this study, Ilwis version-3.4, ArcGIS version-9.3 and ERDAS Imagine Version-9.1 software as well as the statistical software i.e. SPSS version-16 were used to produce the layer maps, which help in the preparation of the malaria-susceptibility index and zones (Rai et al. 2013; Rai and Nathawat 2013). Survey of India (SOI) topographical sheet of 1:50,000 scale is used to digitise the district and development block boundaries. The coordinates of main geo-reference points such as road junctions, malaria-prone areas and current healthcare facilities were measured during field surveys using global positioning system (GPS). During measurement, one receiver served as a base station, while the other collected GPS data at the designated ground control points. To found the relationship between object space and image space, ground control points selected in the model area to conduct all measurements in the National Coordinate System. The various GIS layers or themes developed from the IRS-1C LISS-III remote sensing data of 2008 and Survey of India (SOI) topographical map. A land use map, NDVI and vector layers of water bodies and other important parameters were delineated through ERDAS Imagine version-9.1 and ArcGIS version-9.3 software. A geo-referenced digital map of the development blocks of Varanasi district was used on the GIS platforms.

Three methods are used in this study to detect malaria susceptibility zonation (MSZ) and malaria susceptibility index (MSI). These are

(a) Multiple Linear Regression,
(b) Information Value (Infoval) and
(c) Heuristic Approach

Influential parameters measured in this study for optimum zonation and modeling of study area are: Rainfall (Rf), Temperature (Temp), Population density (Pd), Distance to river (Dri), Distance to road (Dro), Distance to health facilities (Dhf), Land use/Land cover (Lc) and Normalized Difference Vegetation Index (NDVI). These parameters have been applied in all above three statistical methods to create malaria susceptibility zonation (Rai et al. 2013; Rai and Nathawat 2013).

6.2.1 Malaria Inventory Map

A malaria inventory map recognizes definite and feasible areas of existing malaria prevalence and is a simple necessity for MSZ. The malaria inventory map shows the spatial distribution of malaria as points or to scale (Rai et al. 2013; Rai and Nathawat 2013).

Malaria inventory map is often used as the basis for other MSZ techniques or as a basic susceptibility map. Village-wise malaria location data were collected from each primary health centre (PHC) and then the locations were identified using GPS. GPS location data associated to malaria prevalence was introduced into the GIS platform and 500 m buffer zones were created around each point (Fig. 6.1). On the basis of these malaria pixels falling in the study area, the pixels from the whole

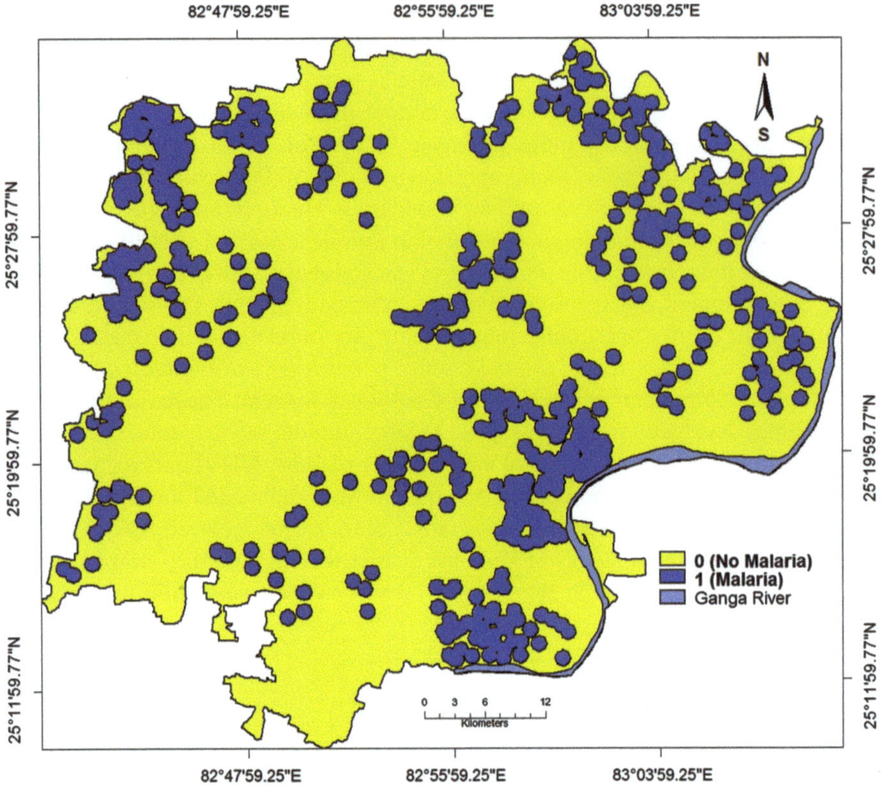

Fig. 6.1 Buffer zones (500 m) around each point location of malaria disease

study area were given one of two values, that is 0 (no malaria pixels) 1 (where malaria pixels are present) (Rai et al. 2013; Rai and Nathawat 2013).

6.3 Malaria Influencing Data Layers and Related Map Preparations

Followings layers are used to produce the malaria-susceptibility model map.

6.3.1 Rainfall

Rainfall is considered to be the utmost vital malaria-triggering parameter causing soil saturation and a rise in pore water pressure. However, there are not many cases of the use of this parameter in stability zonation, possibly due to the difficulty in collecting rainfall data for long periods over large areas (Rai et al. 2012, 2013; Rai and Nathawat 2013).

After interpolation between the amounts of annual rainfall in the study area stations, the isohyets map was created. Finally, this map has been grouped into five classes to prepare the rainfall data layer (Figs. 6.2 and 6.3).

It is confirmed that approximately 57.77 % of the malaria cases occurred in areas with >984 mm rainfall, but in areas with <970 mm rainfall only very low and moderate zones of malaria were found, with only 3.19 % of cases. From this, it can be conclude that increasing amounts of rainfall increases the malaria breeding sources (Table 6.1).

6.3.2 *Temperature*

To take into account the relationship between temperatures and malaria transmission, the temperature data are collected of different time periods (Rai et al. 2012, 2013). The temperature distribution map has been assembled into three main classes i.e. 35.44–35.46 °C, 35.47–35.49 °C and 35.50–35.52 °C (Figs. 6.4 and 6.5). In Table 6.2, it is clearly seen that malaria vectors were highly developed in the 35.44–35.46 °C temperature category in the study area. About 56.50 % of the malaria-prone area pixels were found in this category, whereas only 20.04 % of malaria-prone area pixels were identified in the 35.47–35.49 °C category (Rai et al. 2012, 2013).

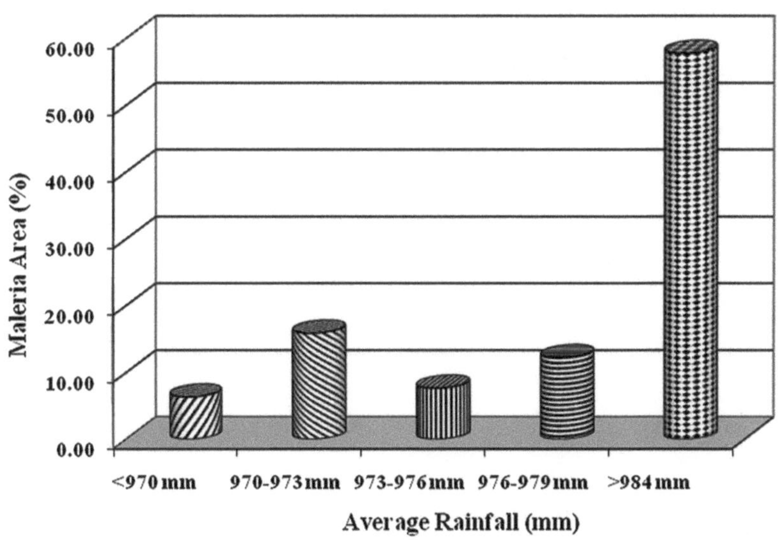

Fig. 6.2 Malaria area (%) vs. average rainfall

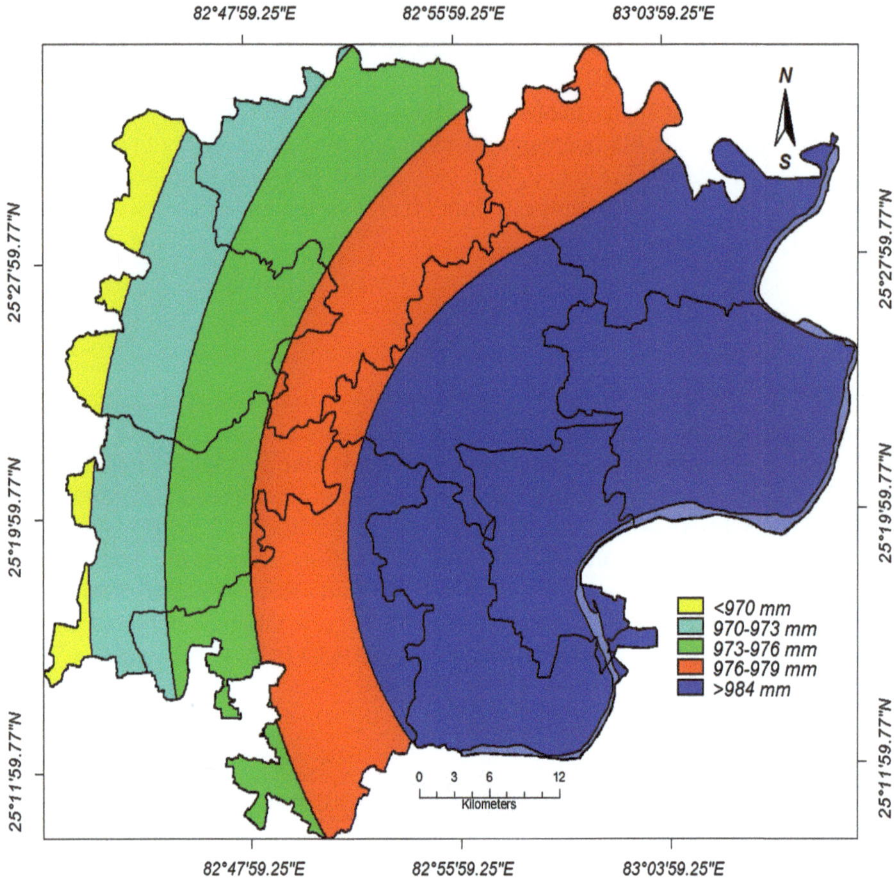

Fig. 6.3 Average rainfall distribution (in mm)

Table 6.1 Malaria area (%) in different rainfall class

Rainfall class	No. of malaria pixel	Total no. of pixel	Malaria area (%)
<970	7897	19,519	6.38
970–973	19,682	68,937	15.89
973–976	9542	99,487	7.70
976–979	15,175	143,285	12.25
>984	71,547	280,162	57.77

6.3.3 Population Density

The overall population distribution in the district is meticulously connected to the physical and sociocultural factors. Population distribution is a dynamic process which manifests the variable nature of man's adjustment to physical resources. Population density has been encountered under numerous typological purviews to

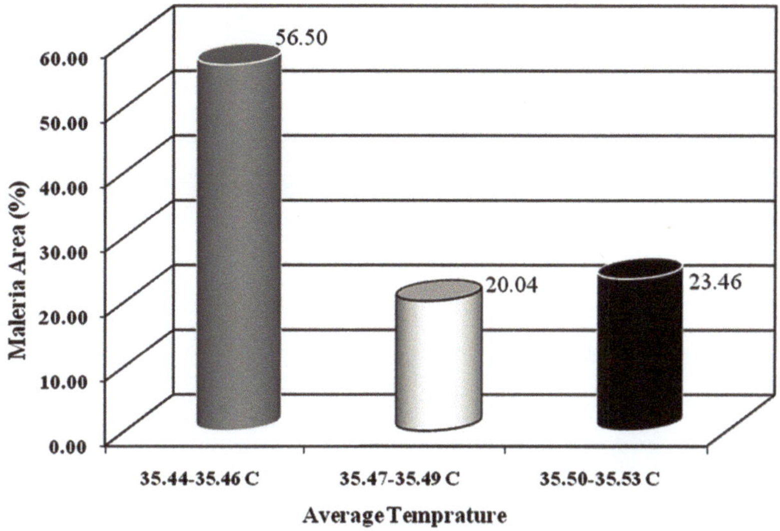

Fig. 6.4 Malaria area (%) vs. average temperature

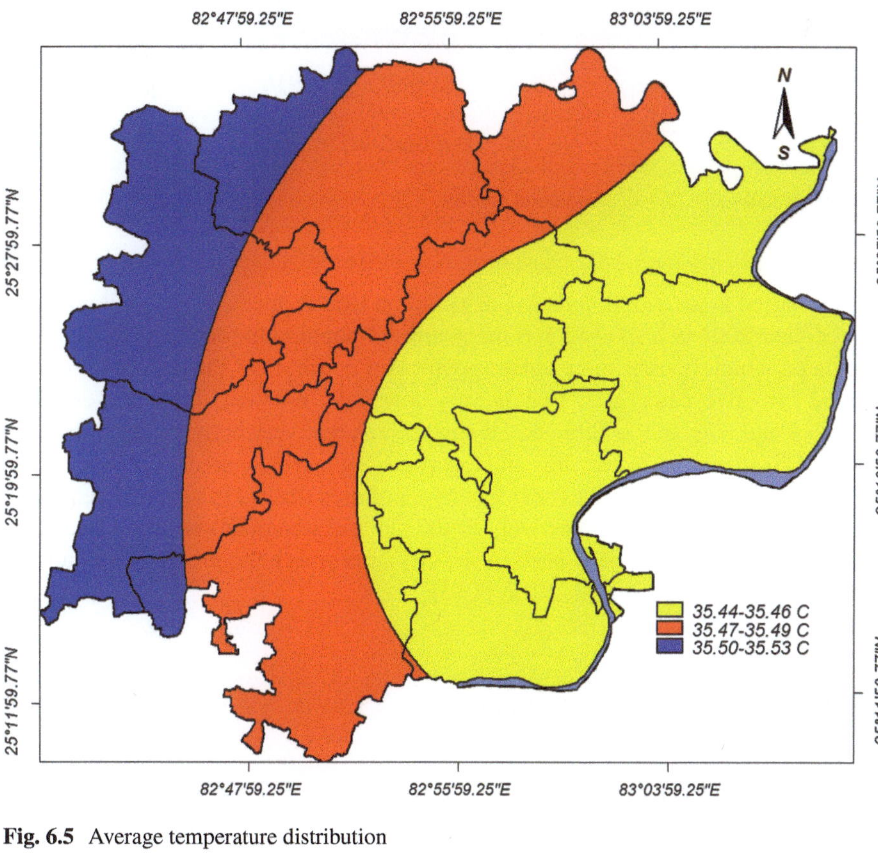

Fig. 6.5 Average temperature distribution

Table 6.2 Malaria area (%) in different temperature class

Temperature class	No. of malaria pixel	Total no. of pixel	Malaria area (%)
35.44–35.46 °C	69,974	266,938	56.50
35.47–35.49 °C	24,812	234,152	20.04
35.50–35.53 °C	29,057	110,300	23.46

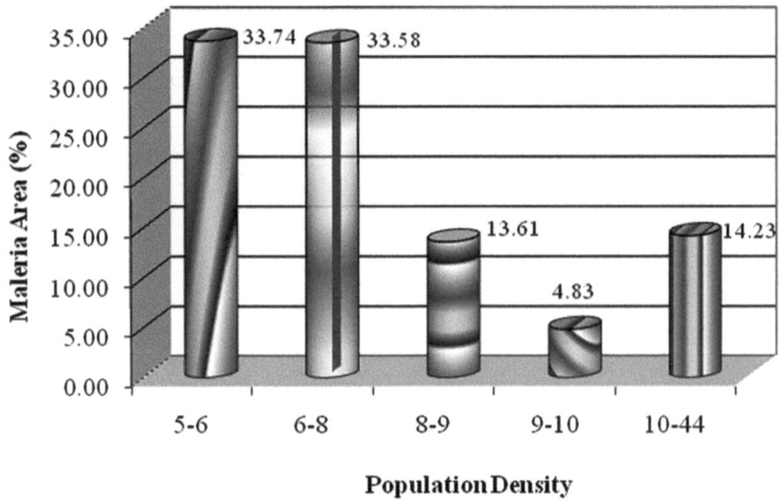

Fig. 6.6 Malaria area (%) vs. population density

divulge varied aspects of population distribution (Rai et al. 2012, 2013). Population data of year 2001 is also used and the projected population for the year 2010 was calculated which is used to calculate population density. On the basis of that area divided into five categories, that is very low, low, medium, high or very high (Figs. 6.6 and 6.7). In the Table 6.3, it can be seen that very high malaria incidence (33.74 %) is identified in very low and low population density areas especially in the rural parts of Varanasi district while where population density is high and very high, only 4.83 % and 14.23 % respectively of the malaria area pixels are found. Most of the very high malaria zone is seen in the city area where the projected population density is very high (Rai et al. 2012, 2013).

6.3.4 *Distance to River/Streams*

Distance to rivers/streams is one of most significant parameters that contribute to an increase in malaria parasites and malaria disease in the area. The proximity of the populated area to drainage structures is a vital parameter for malaria vector

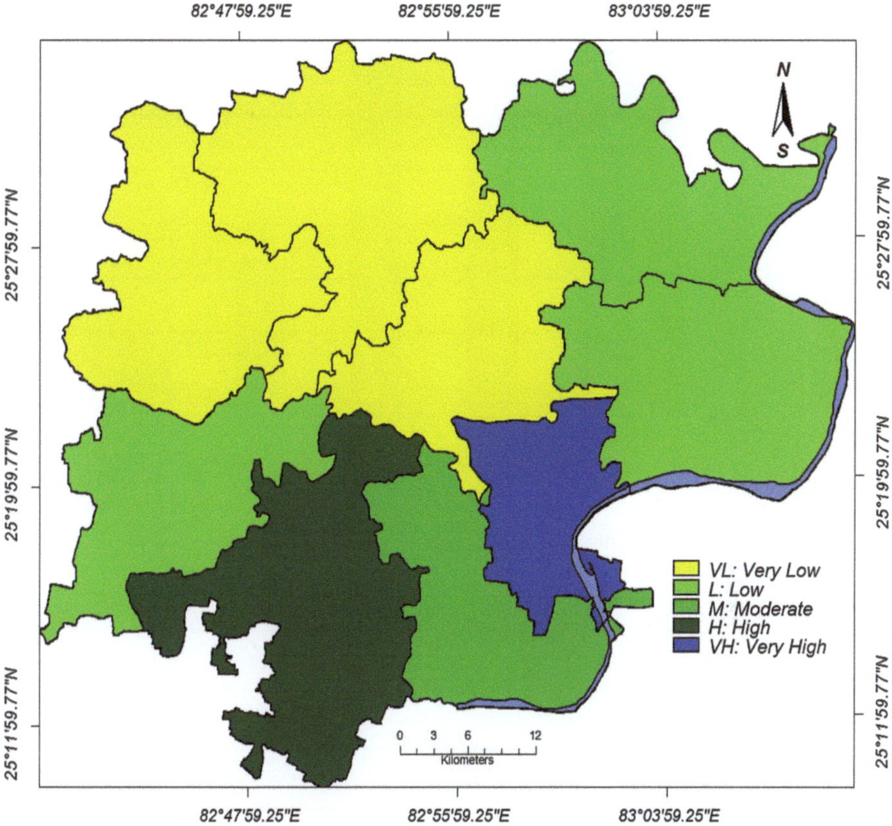

Fig. 6.7 Population density

Table 6.3 Malaria area (%) in different population density class

Population density	No. of malaria pixel	Total no. of pixel	Malaria area (%)
Very low	41,787	222,199	33.74
Low	41,584	212,569	33.58
Moderate	16,861	53,352	13.61
High	5985	84,805	4.83
Very high	17,626	38,460	14.23

breeding sources. Streams may also inauspiciously affect those in low-lying areas, specially villages and settlement adjoining the Varuna river (Rai et al. 2012, 2013). A thorough field investigation was accomplished to determine the effects of rivers/streams on malaria prevalence (Figs. 6.8 and 6.9). The malaria area percentage in each buffer zone is given in Table 6.4. This shows that 23.23 % of the malaria area is closely located within the <1000 m buffer zone. It is recognised that about 31.32 % of the malaria area pixels are come in the buffer zone of 3000–6000 m. and about 3.56 % of the malaria area pixels are found >10,000 m from a river/stream and only

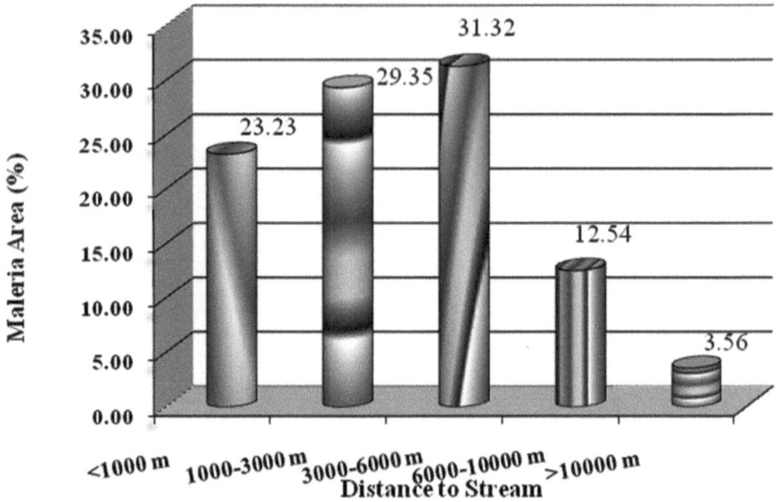

Fig. 6.8 Malaria area (%) vs. distance to streams

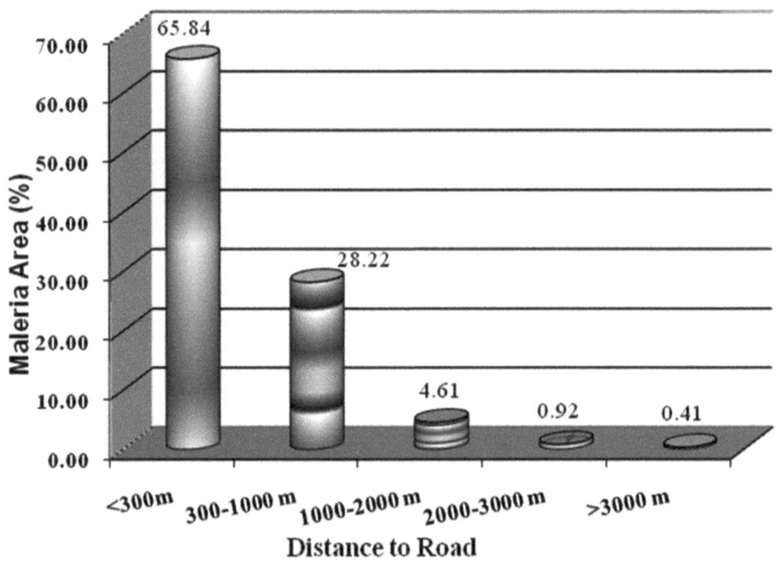

Fig. 6.9 Malaria area (%) vs. distance to road

very low and moderate categories are available in this zone, which is usually because of the influence of some of the other indicators/variables, so at this distance malaria indicators or breeding sources have little effect on persons living in the area (Rai et al. 2012, 2013).

Table 6.4 Malaria area (%) vs. distance to streams

Distance class	No. of malaria pixel	Total no. of pixel	Malaria area (%)
<1000 m	28,766	132,325	23.23
1000–3000 m	36,354	173,300	29.35
3000–6000 m	38,783	184,054	31.32
6000–10,000 m	15,533	80,756	12.54
>10,000 m	4407	40,955	3.56

The vital outcome found in this study is that as the distance to rivers/streams increases, the percentage of malaria-affected area pixels decreases (Rai et al. 2012, 2013).

One important thing it is also found in this study that as the distance to rivers/stream increases, percentage of malaria effected area pixels are decreasing.

6.3.5 *Distance to Road*

The distance to roads is also a significant factor as it can be used as an estimation of the access to present healthcare facilities in the area (Rai et al. 2012, 2013). Buffer zones are calculated on the path of the road to determine the effect of the road on malaria prevalence (Figs. 6.9 and 6.10). The malaria area pixel percentage in each buffer zone is given in the Table 6.5 shows that 65.84 % of the malaria area pixels are closely located within the <300 m buffer category, although only a very nominal 0.92 % of the malaria area is seen in the buffer zone of 2000–3000 m. Only 0.41 % of malaria pixels are located in the >3000 m buffer category. Here, it can be very-well identified that as the distance to roads increases, the malaria area percentage shows a decreasing trend.

6.3.6 *Distance to Health Facilities*

Health facilities of Varanasi district are depended on mostly modern allopathic of treatment. To know the distributional pattern of health care facilities, data has been collected from chief medical officer (CMO) office and government hospital situated in rural areas of Varanasi. The existing health facilities both in rural and urban area are identified through GPS. There are various classes of health centre providing infrastructure and treatment in the district (Rai et al. 2012, 2013). The PHC's are spread in the district located at an interval of 10–20 kms. and the tahsil hospitals are located about 50 km apart. The hierarchical with the distribution of medical centre's of the district allows a close relationship with the hierarchy a central places and population size of the settlement. Moreover, the transport network has also influenced the development of health care facilities (Rai et al. 2012, 2013). Percentage

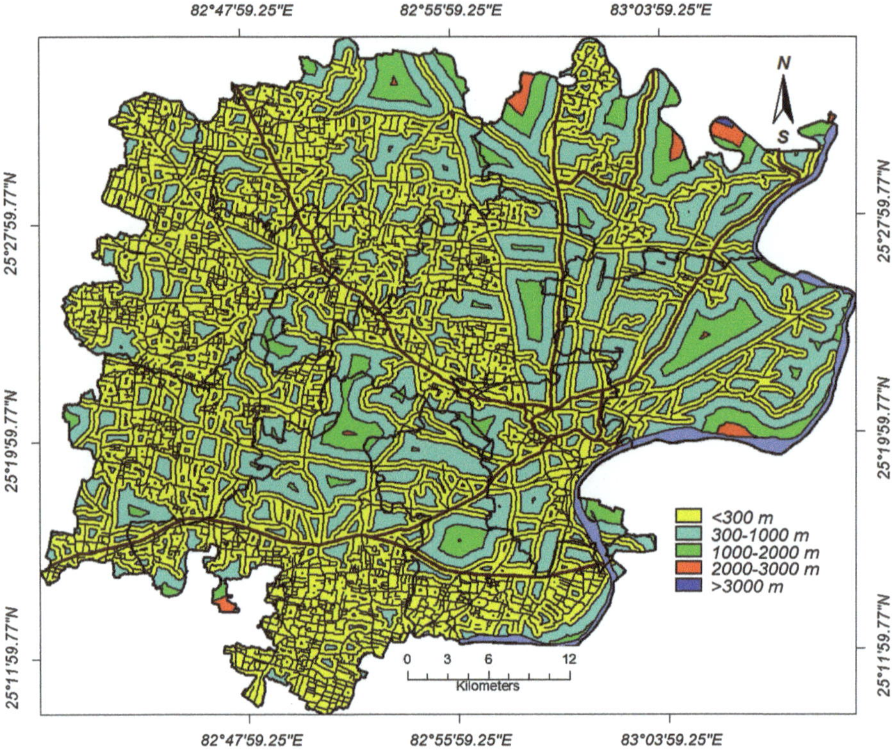

Fig. 6.10 Distance to road network

Table 6.5 Malaria area (%) vs. distance to road

Distance class	No. of malaria pixel	Total no. of pixel	Malaria area (%)
<300 m	81,541	390,258	65.84
300–1000 m	34,954	177,295	28.22
1000–2000 m	5706	38,918	4.61
2000–3000 m	1134	3499	0.92
>3000 m	508	1420	0.41

of malaria area pixels is very much associated to distance to health facilities (Figs. 6.11 and 6.12). In Table 6.6, it is clearly seen that, 6.59 % of malaria area is belonging to 0–1000 m buffer distance to health facilities and maximum about 35.14 % of malaria area identify in 3000–6000 m buffer distance.

Table 6.6 displays that as the distance to health facilities increases, malaria area are also increasing, except in >10,000 m buffer zones (8.94 % malaria area pixels are available). In this area may be malaria breeding sources are not well developed as much than the other area.

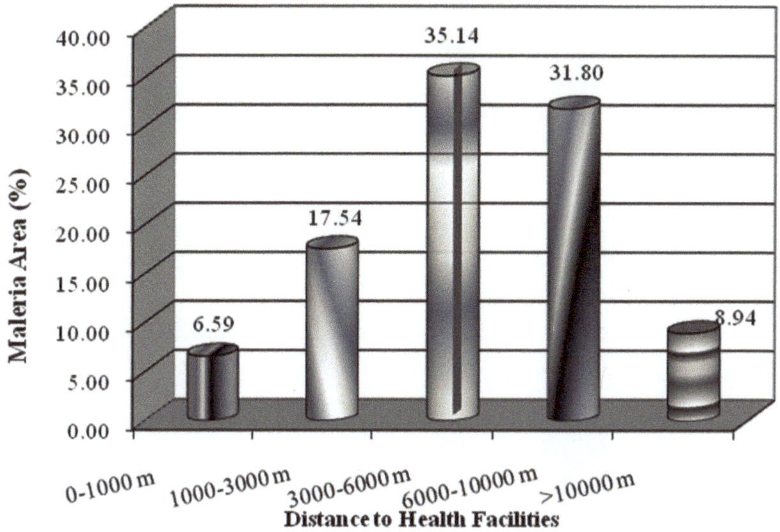

Fig. 6.11 Malaria area (%) vs. distance to health facilities

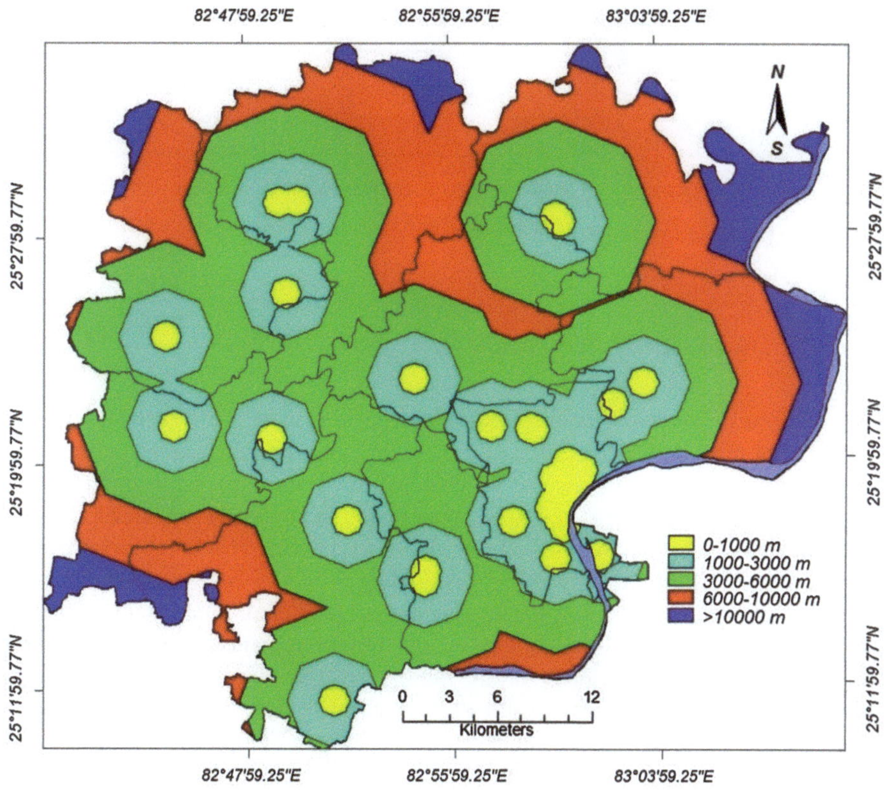

Fig. 6.12 Distance to health facilities

Table 6.6 Malaria area (%) vs. distance to health facilities

Distance class	No. of malaria pixel	Total no. of pixel	Malaria area (%)
0–1000 m	8158	29,136	6.59
1000–3000 m	21,720	145,065	17.54
3000–6000 m	43,515	242,719	35.14
6000–10,000 m	39,376	147,357	31.80
>10,000 m	11,074	47,113	8.94

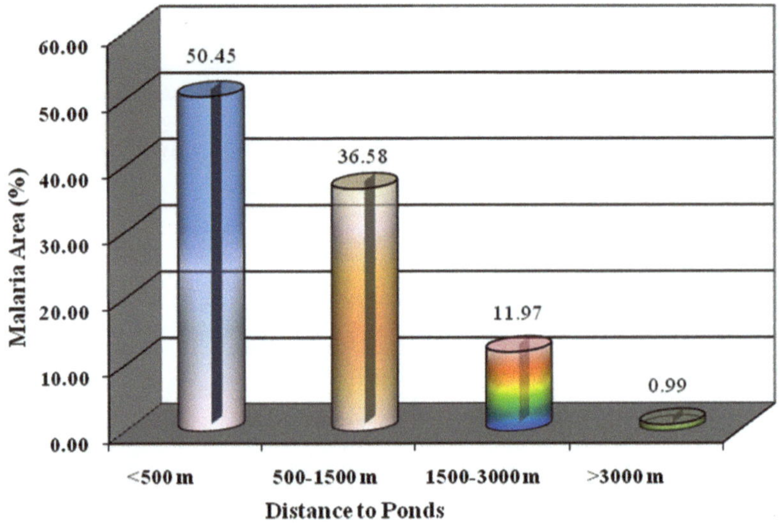

Fig. 6.13 Malaria area vs. distance to ponds

6.3.7 *Distance to Water Ponds*

The locations of water ponds in the study area are extracted from IRS-1C LISS III data of 2008. Five buffer zones were developed along the water ponds to determine the consequence of the distance to water ponds on malaria prevalence (Figs. 6.13 and 6.14). In Varanasi, there used to be many ponds and tanks dating back to earliest times. In addition to helping as the holy places for holding Hindu religious rituals, they also played a noteworthy role in rainwater collection and thus served as an important sources for ground water replenishment (Rai et al. 2012, 2013). Though, due to the rapid increase of the population, most of these ponds have been wiped off the map of Varanasi completely or are rapidly deteriorating. The main source of pollution in the ponds is loads of solid waste or garbage. The solid and liquid wastes generated from household and industrial activities are dumped and released into uncontrolled sites. These sites leak into the low-lying areas where the tanks and ponds are situated and due to this malaria vectors develop very effortlessly and

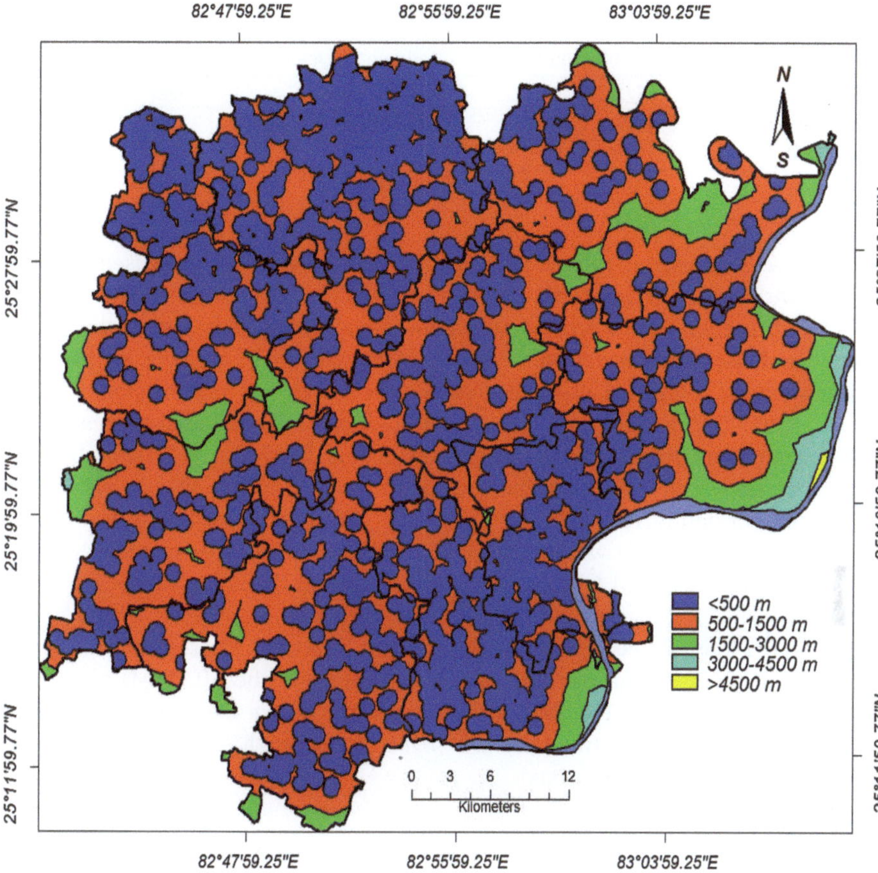

Fig. 6.14 Distance to ponds/water bodies

many cases of malaria are identified near to these contaminated ponds. It is seen that about 50.45 % of malaria area pixels occurred in the <500 m buffer category of ponds and only 0.99 % of malaria area pixels were found in the zone of >3000 m buffer category of ponds. Table 6.7 reveals that as the distance to ponds or water bodies increases, the percentage of malaria area pixels decreases.

6.3.8 *Normalized Difference Vegetation Index (NDVI)*

Vegetation is frequently connected with vector breeding, feeding and resting locations. Because malaria is vector-borne, there are numerous remotely sensed abiotic and biotic environmental factors that are pertinent to the study of malarial transmission and habitat places of the vector (Rai et al. 2012, 2013). A number of vegetation

Table 6.7 Malaria area (%) vs. distance to ponds

Distance class	No. of malaria pixel	Total no. of pixel	Malaria area (%)
<500 m	67,032	270,681	50.45
500–1500 m	48,608	283,869	36.58
1500–3000 m	15,909	39,612	11.97
>3000 m	1319	17,228	0.99

indices have been used in remote sensing, but the utmost extensively used index to augment the vegetation areas and crop fields is NDVI. NDVI values range from −1 to +1, with higher values representing denser vegetation. The higher the NDVI value, the denser the vegetation. Many diseases and their causative agents have environmentally connected attributes that must be present for transmission or infection to occur. NDVI and remotely sensed variables provide additional approaches of exploring and well defining these attributes (Rai et al. 2012, 2013). Distributions of diseases related with arthropod and gastropod vectors, classifiable as whichever intermediate or definitive hosts depending upon the presence or absence of sexual reproduction of the agent while hosted, are defined by their topographies such as land use land cover (LULC) and proximity to aquatic habitats (Rai et al. 2012, 2013).

The NDVI map has been assembled into three important classes and it is found that about 19.39 % of malaria area pixels are found in the –0.288–0 categories (Figs. 6.15 and 6.16). In the Table 6.8, it is clearly shown that about 80.69 % of malaria area pixels are identified in the 0–0.986 category, which is the class of agriculture, vegetation and fallow land.

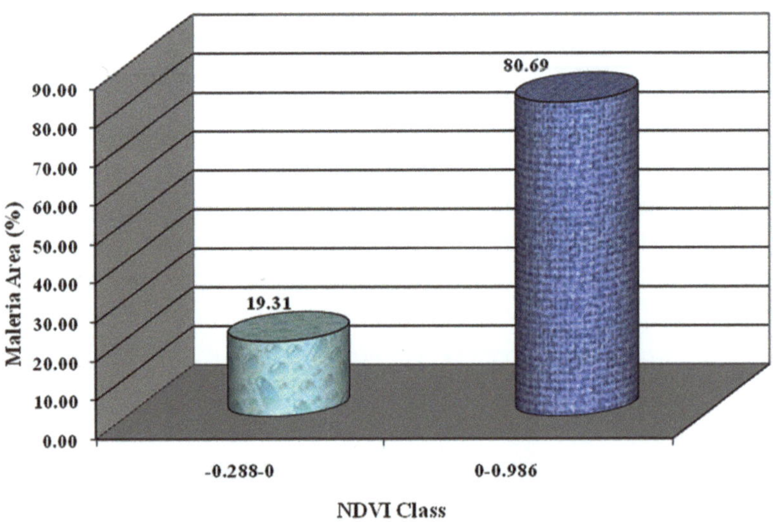

Fig. 6.15 Malaria area (%) vs. NDVI

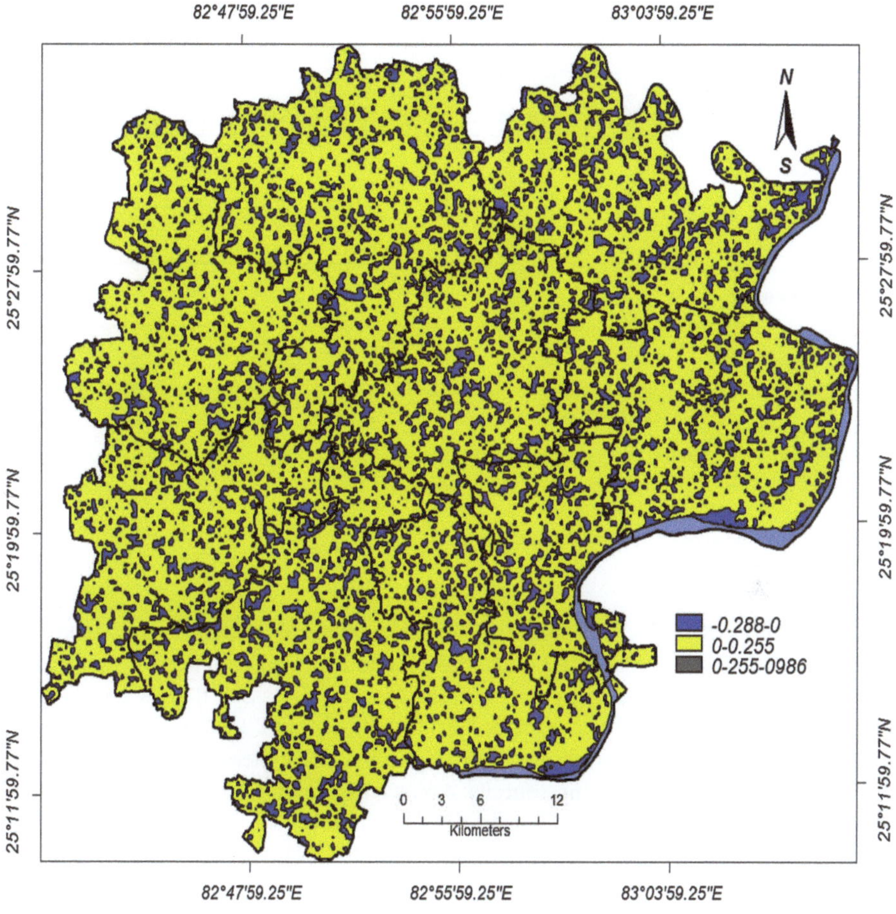

Fig. 6.16 Normalized Difference Vegetation Index (NDVI)

Table 6.8 Malaria area (%) vs. NDVI

Distance class	No. of malaria pixel	Total no. of pixel	Malaria area (%)
−0.288–0	23,909	115,299	19.31
0–0986	99,934	496,053	80.69

6.3.9 Land Use Land Cover (LULC)

Land use land cover (LULC) information is also very important parameters to estimate malaria susceptibility map and compute malaria susceptibility zone using multi linear regression model (Rai et al. 2012, 2013; Rai and Nathawat 2013). The

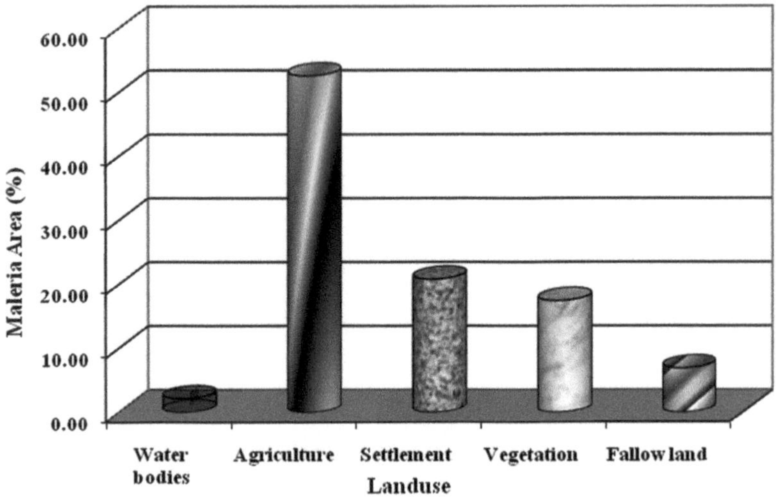

Fig. 6.17 Malaria area (%) vs. land use class

land use land cover (LULC) map of the study area has been prepared from the IRS-1C LISS III remote-sensing data from 2008. The land use map is prepared in the image processing platform to highlight five main classes that is agricultural fields, settlement, vegetation, water bodies and fallow land. In this study we found that agriculture and vegetation are very important parameters and play important roles as malaria vector breeding sources. Areas which include dense vegetation provide favourable conditions for malaria vectors. The presence of crop fields, particularly in those areas where rice cultivation is leading, is also critical as a malaria vector breeding source. Many parts of Varanasi district have fertile crop fields and where irrigation facilities are quite good; farmers promote rice (Rai et al. 2013; Rai and Nathawat 2013).

This area, with good agriculture fields, should be prone to the occurrence of malaria in certain cases (Figs. 6.17 and 6.18). By using land use land cover and malaria maps to identify the distribution of malaria disease, it is clearly seen that about 52.64 % of malaria pretentious pixels out of the total pixels occurred in the agriculture class whereas 20.84 %, 17.44 %, 6.90 % and 2.17 % of the malaria area pixels occurred in the settlement, vegetation, fallow land and water bodies classes respectively (Table 6.9).

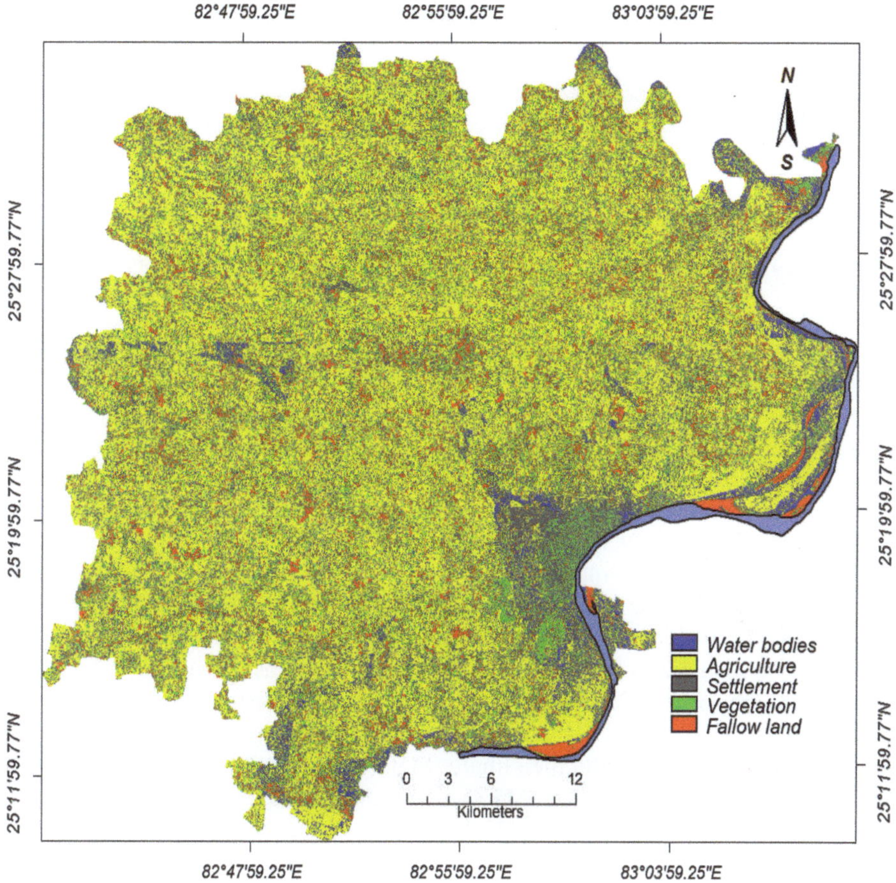

Fig. 6.18 Land use/land cover classes

Table 6.9 Malaria area (%) vs. land use class

Land use class	No. of malaria pixel	Total no. of pixel	Malaria area (%)
Water bodies	2690	19,282	2.17
Agriculture	65,188	330,581	52.64
Settlement	25,810	114,660	20.84
Vegetation	21,595	99,812	17.44
Fallow land	8544	46,601	6.90

6.4 Methods and Results of Statistical Methods

Three statistical methods i.e. Multiple Linear Regression, Information Value (Infoval) and Heuristic Method.are used in this study to develop malaria susceptibility index (MSI) and malaria susceptibility zonation (MSZ)

6.4.1 Multiple Linear Regression Method (Step-Wise Method)

Multivariate statistical analyses of causative features controlling malaria disease occurrence may indicate the relative contribution of each of these factors to the degree of disease occurrence within a well-defined land unit. These analyses are dependent on the presence or absence of stability phenomena within the units (Van Westen 1993; Rai et al. 2012). In order to carry out multivariate analysis of data and to determine the all parameters responsible for malaria disease in Varanasi district, a multiple linear regression method is used. Multiple Linear regression models is constructed for malaria cases stated in the study area, as the dependent variable and many time based groupings of temperature, rainfall and NDVI data as the independent variables (Rai et al. 2012). The multiple linear regression method described that how the susceptibility of malaria as the standard deviation of independent variables and interpreters change. All these related parameters are calculated in SPSS statistical software using multiple linear regression method and crossed to each other and then finally Malaria Susceptibility Index (MSI) and Malaria Susceptibility Zonation (MSZ) are produced (Rai et al. 2012).

In this study equation of the theoretical model will be described as follows.

$$M = B_0 + b_1 X_1 + b_2 X_2 + b_3 X_3 + \ldots + b_m X_m + \Sigma \quad (6.1a)$$

Where, M is the existence of Malaria in each unit, X's are the input independent variables (or instability parameters) perceived for each mapping unit, the B's are coefficients assessed from the data through statistical methods, and ε signifies the model error (Irigaray et al 2007; Rai et al. 2012). Using all the above parameters, Malaria Susceptibility Index and Malaria Susceptibility Zones are developed. All these indicators are very useful to create a relationship between in malaria breeding source.

The following GIS processes are normally used in multivariate statistics for malaria susceptibility zonation:

(a) Determination of the list of factors that will be comprised in the analysis. As many input maps are of alphanumeric type, they should be generally converted into numerical maps. These maps can be converted into several maps with presence/absence values for each land unit, or one map with values as percentage cover of each parameter class or expert values according to increasing observed malaria.

(b) Change the observed malaria map into numerical map by attributing the value 1 to observed malaria areas and the value zero to the other areas.
(c) Export for each pixel of the map the different numerical values to a statistical package for succeeding analysis.
(d) Import the predictable outcomes into a raster map.
(e) Classification of the map into susceptibility classes.

In this study, there are ten causative factors for the multiple linear regression model of malaria susceptibility as independent variables, all variables are numerical. Only one variable i.e. land use is alphanumeric type or classes that cannot be processed directly.

In order to process the multiple linear regression model of malaria susceptibility with the current data, the SPSS-16 software is used to process the data to evaluate the regression.

The standardized regression coefficients or t-values, which are the regression coefficients expressed in units of their standard deviation, designate the relative contribution of every causative factor to the occurrence of malarias (Davis 1986) and enable to choose whether a certain factor has a substantial contribution in the equation. If the absolute t-value is larger than 1.96, the coefficient can be considered significant with 95 % confidence.

The values given in Appendix-1 designate that causal factors such as rainfall, temperature, distance to hospital, distance to health centers, distance to streams/river, distance to road, distance to ponds/water bodies, population density, NDVI play an important influence in the regression model and subsequently have an important influence on malaria susceptibility. Landuse is less significant. The land use land cover (LU/LC) of the regression test when the variable into the equation for the amplitude is increased to 5 % significance level test. So that land use variable is stopped and removed from the model.

Also the following statistics are used to assess the overall outcome of the model: R2 = 0.065 gives the amount of variance accounted for by the model.

Hence, only 6.5 % of the variability in detected malarias is elucidated by the model, which is a rather poor result.

MSE = 0. 389 is mean standard error of estimate, i.e. the square root of the residual mean square error which measures the unexplained variability in the dependent variable. In view of the fact that model produces M values between 0 and about 0.1, the estimated standard error is rather high.

The malaria model equation used for calculation in MSI and MSZ in multiple linear regression method is:

$$M = (145.54) + (0.009 * Pd) + ((2.572) * Dh) + ((-9.406) * Dwo) + \\ ((-1.354) * Ds) + ((-1.009) * Dhc + ((-0.023) * Rf) + ((-3.47) * At) + \\ ((-1.98) * Dro) + ((-0.106) * NDVI)$$

Where, M is the occurrence of malaria in each unit.

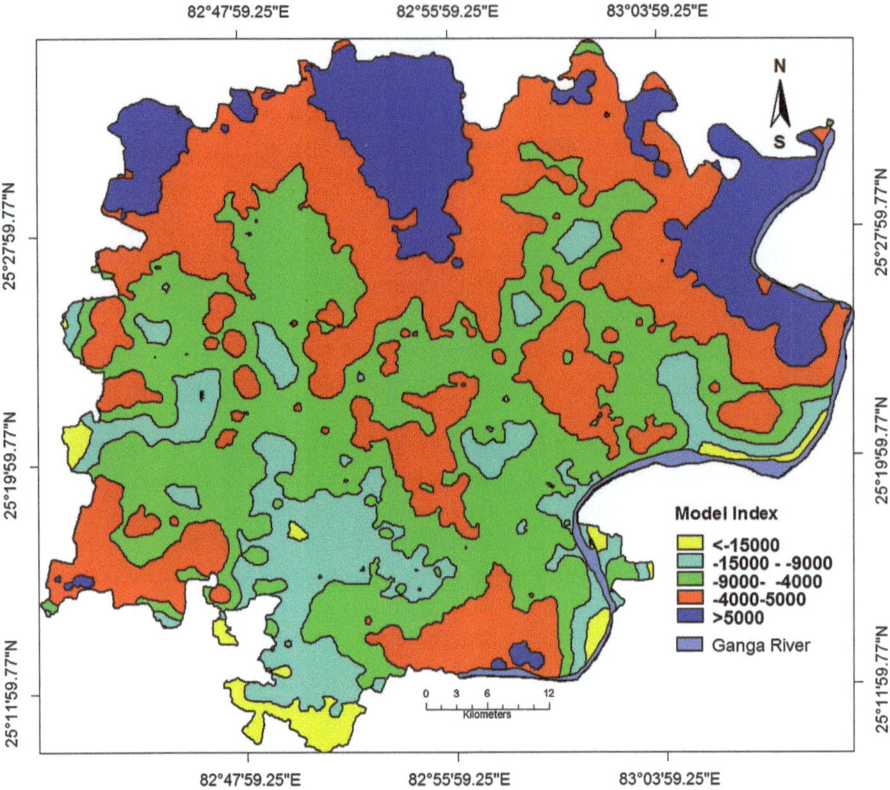

Fig. 6.19 Malaria Susceptibility Index (MSI) based on multiple linear regression method

The Malaria Susceptibility Index (MSI) values from the multiple linear regression method are found to lie in the range from < −15,000 to >5000 (Fig. 6.19).

In the Figs. 6.19 and 6.21, it is very well seen that about 36.06 % of the pixel area exist in −4000–5000 model index class and susceptibility index < −15,000 contain 2.57 % of the pixels area while about 13.43 % of the pixels area comes in >7000 susceptibility index class. The cumulative frequency curve of MSI values has been segmented into five classes representing near equal distribution to yield five malaria susceptibility zones (MSZ) i.e. very low, low, moderate, high and very high (Table 6.10, Figs. 6.20 and 6.22). Figure 6.22 shows that 45.40 % of malaria area falls under high susceptibility level whereas only 0.36 % and 6.66 % of malaria goes to very low and low susceptibility level.

Among the variables entered in the model based on the standardized Beta values in order of preference are: Pd (with a beta coefficient of 0.211), distance to hospitals, Dh (with a beta coefficient of (with a beta coefficient of 0.196), distance to ponds/water bodies, Dwb (with a beta coefficient of −.154), distance to stream/river, Ds (with a beta coefficient of −.110), distance to health centers, Dhc (with a beta

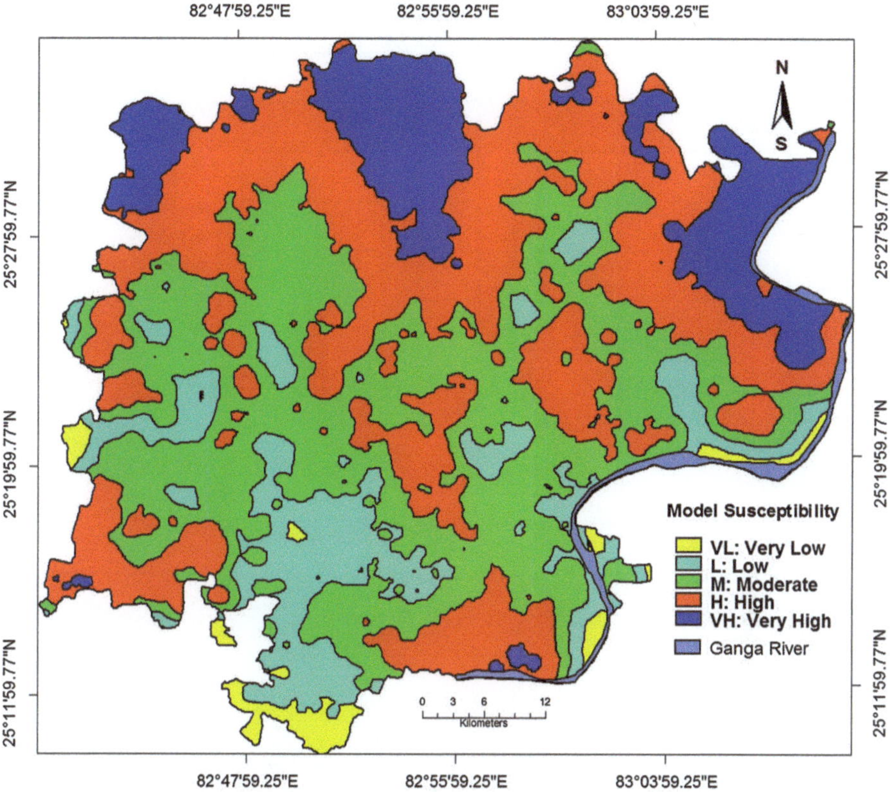

Fig. 6.20 Malaria Susceptibility Zone (MSZ) based on multiple linear regression method

Table 6.10 Status of malaria area percentage vs. malaria level based on the multiple linear regression method

Malaria level	Total no. of pixels	Pixel (%)	Malaria area (sq.km)	Malaria area (%)
Very low	15,703	2.57	39.2575	0.36
Low	78,946	12.92	197.365	6.66
Moderate	214,138	35.03	535.345	27.69
High	220,406	36.06	551.015	45.40
Very high	82,066	13.43	205.165	19.87
Total	611,259	100.00	1528.1475	100.00

coefficient of −.043), rainfall, Rf (with a beta coefficient of −.215), average temperature, At (with a beta coefficient of −.181), distance to road, Dro (−.020), NDVI (−.015).

Among the independent variables, distance from the streams, distance from the road and also because of rainfall, low beta levels are knowingly higher than 5 % out

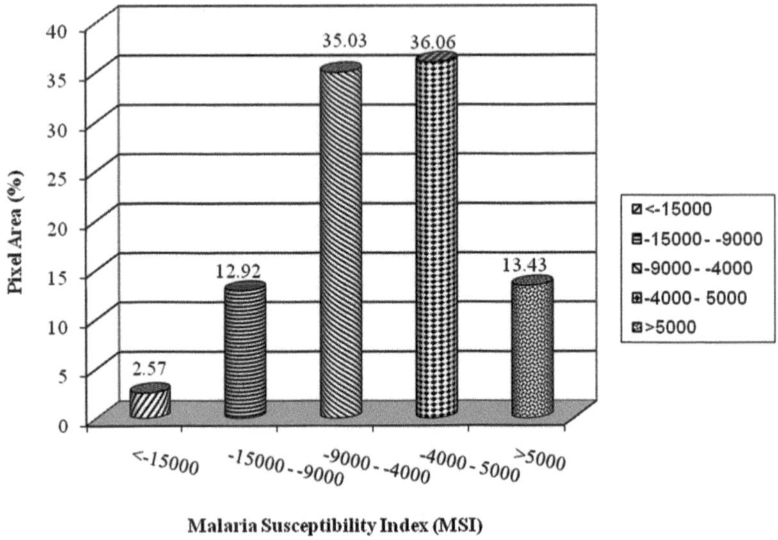

Fig. 6.21 Distribution frequency histogram of malaria based on the multiple linear regression model

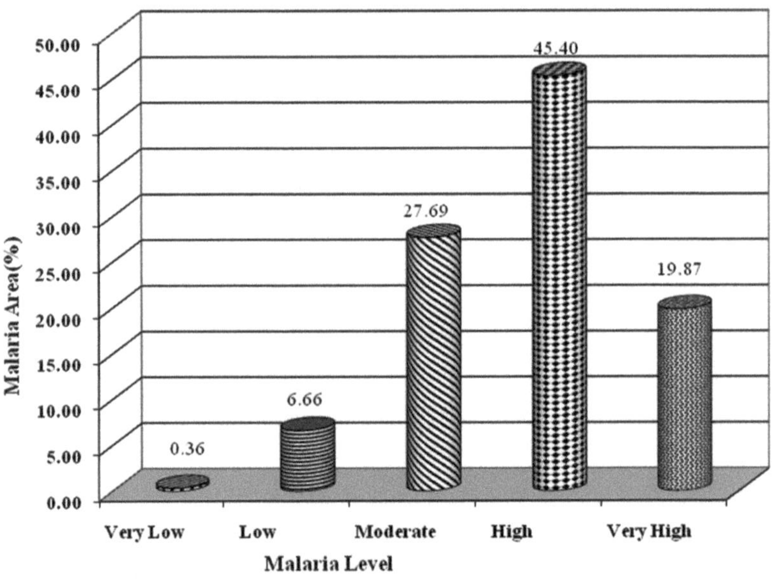

Fig. 6.22 Malaria area vs. malaria susceptibility level based on multiple linear regression method

of the equation (in the statistical calculations completed in this manner will be attached to all). Then, calculated coefficients are used in the matrix of dataset and the equation is envisioned for all of the 38,622×9 sample pixels of the study area (Rai et al. 2012, 2013; Rai and Nathawat 2013). Finally column of the equation outcome for analyzing and creating malaria susceptibility map has been moved into GIS software (ILWIS 3.4).

6.4.2 Information Value Method (InfoVal)

The InfoVal method for MSZ deliberates the probability of malaria occurrence within a certain area of every class of a thematic (Shah 2007; Rai et al. 2013). In this model, weights of a specific class in a thematic are identified as

$$W_i = \ln\left(\frac{Densclas}{Densmap}\right) = \ln\frac{Npix(S_i)/Npix(N_i)}{\sum_{i=1}^{n} Npix(S_i)/\sum_{i=1}^{n} Npix(N_i)} \tag{6.1b}$$

where W_i is the weight specified to the ith class of a specific thematic layer, *Densclas* is the malaria density within the thematic class, *Densmap* is the malaria density within the whole thematic layer, $Npix(S_i)$is the number of malaria pixels in a certain thematic class, $Npix(N_i)$the total number of pixels in a certain thematic class and N is the number of classes in a thematic map. The natural logarithm is used to take care of the large difference in the weights. Thus, the weight is considered for various classes in each thematic. The thematic is overlaid and added to make a MSI map (Rai et al. 2013). For near-equal subdivision of the MSI, the cumulative frequency curve is characterised into five zones based on malaria susceptibility (that is very high, high, moderate, low and very low). Information analysis encompasses two precise steps, i.e. bivariate analysis and multivariate analysis (Fig. 6.23).

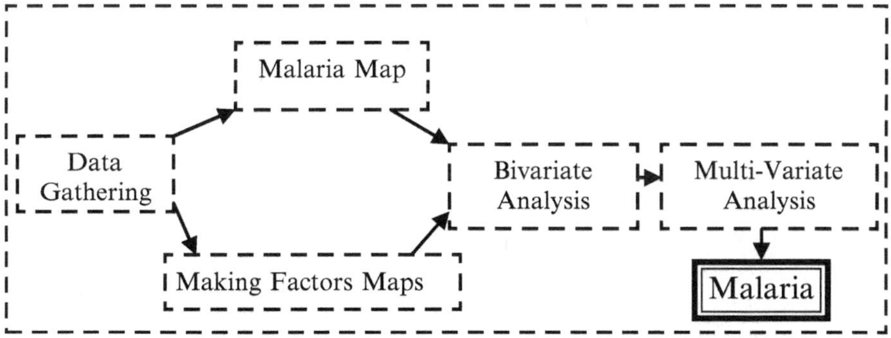

Fig. 6.23 Different stages of making malaria susceptibility zonation map in information value method

6.4.2.1 Bivariate Analyze

Determining relationship between affective parameters and their sub-group with malaria is premeditated by assessing relative area occupied by malaria in the study area and the area of any sub-group. In this analyze, two ratios are used:

(a) *P* is the ratio of malaria area and study area

$$P = \frac{M}{N} \tag{6.2}$$

where *M* is the number of the malaria cases in the study area and *N* is whole of the study area

(b) *Pi* is the ratio of malaria area in the discrete parameters:

$$Pi = \frac{Li}{Ni} \tag{6.3}$$

where, *Mi* is the number of the malaria disease cases into the *i*th variable and *Ni* is the study area including the *i*th variable.

Then the relation between these two values, the *Pi*/*P* information value, can be calculated from the *i*th variable in the predicted malaria potential characterised with *Mi*:

$$Mi = \frac{Pi}{P} = \frac{Mi/Ni}{M/N} \tag{6.4}$$

Then for each variable and its sub-group, the *Mn* of each *Mi*, which designates positive and negative zones, has been calculated. If the calculated *Mn* are positive, this specifies that the pixels including the *i*th variable have a greater incidence of malaria than mean of the study area. This shows the susceptibility of these parameters to instability. Negative values stipulate stability which means that there is no existence of malaria pixels (Rai et al. 2013).

6.4.2.2 Multivariate Analysis

After producing thematic maps by interpolating the outcomes of each variable information value, maps are divided into sample areas of 200 × 200 pixels (Rai et al. 2013). Then the numerical values results, which comprised 38,622 samples, have been transferred to MS- excel software and the final information value has been calculated as follows:

$$Ij = \sum xji \cdot Ii = \sum xji \cdot \frac{Mi / Ni}{M / N} \tag{6.5}$$

where $j = 1, 2, \ldots, n$ designate the number (area) of networks, $i = 1, 2, \ldots, n$ indicate the number of variables, xji is the quantity of the ith variable in the jth indicator, where $i = 1$ means malaria is present $i = 0$ indicates no malaria, and Ii is the information values resulting from the ith variable (Rai et al. 2013). After statistical calculation of the model, information resulting from model has been transferred to GIS (ILWIS 3.4) and the MSI map has been developed (Fig. 6.24).

The next step in this process is assessing the quantities of crucial information values and isolating the degree of susceptibility, which are usually based on calculated values (Rai et al. 2013). For the malaria, the crucial quantity can be certain as the quantity by which the frequency of malaria in the higher quantities is high. To calculate this amount, the information values calculated from malaria area, distribution frequency histogram has been used (Fig. 6.26).

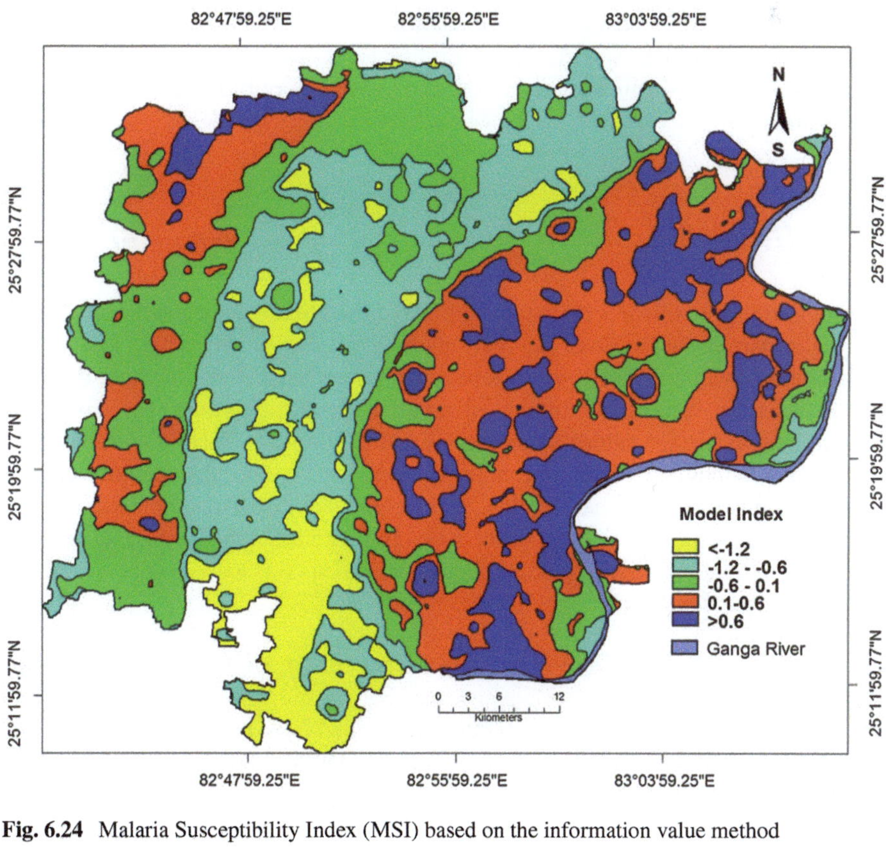

Fig. 6.24 Malaria Susceptibility Index (MSI) based on the information value method

Table 6.11 Status of malaria area percentage vs. malaria level based on the information value method (InfoVal)

Malaria level	Total no. of pixels	Pixel (%)	Malaria area (sq.km)	Malaria area (%)
Very low	58,782	9.61	47.95	3.87
Low	143,609	23.49	149.61	12.08
Moderate	149,494	24.45	221.53	17.89
High	181,759	29.73	493.70	39.86
Very high	77,746	12.72	325.64	26.29
Total	611,390	100	1238.43	100

Information Value Method (InfoVal)

As shown in this histogram, distribution of malaria area in the information value = 0.1–0.6 is looking sensitive, 29.73 % of pixels area have the quantities more than this amount, so this value can be certain as vital value for malaria (Rai et al. 2013). Pixels networks having information value more than 0.1 based on malaria area percentage have been separated into two groups i.e. high and very high susceptible and pixels having lower than 0.1 information value have been divided into three level, low, very low and moderate susceptibility (Table 6.11 and Fig. 6.25). Table 6.11 and Fig. 6.27 also emphasized that, about 3.87 % of the malaria area comes under very low malaria susceptibility level whereas about 39.86 % and 26.29 % of the malaria area come in high and very susceptibility level respectively (Rai et al. 2013).

6.4.3 *Heuristic Approach (Qualitative Map Combination)*

The heuristic approach or weightage method is based on the expert opinion and the relative importance of various causative parameters derived from field knowledge (Rai and Nathawat 2013). The various data layers i.e. Distance to hospitals, distance to health centers, distance to streams, distance to ponds, rainfall etc. etc. have been decided in a weightage values (from 1 to 5) and similarly, each class within a layer has given a weightage values, the highest class has 5 value, the medium class 3 value and the lowest class has 1 value. The weights are given to the classes of each thematic layer respectively, to produce weighted thematic maps, which have been overlaid and mathematically added according to Eq. 6.6 to calculate a Malaria Susceptibility Index (MSI) map (Rai and Nathawat 2013).

$$MSI = Pd + Rf + Dri + Dpo + Dhf + Dro + Temp + Lu / Lc + NDVI \quad (6.6)$$

where *Pd, Rf, Dri, Dpo, Dhf, Dro, Temp., Lu/Lc and NDVI* are distribution-derived weights for population density, rainfall, distance to river/streams, distance to ponds,

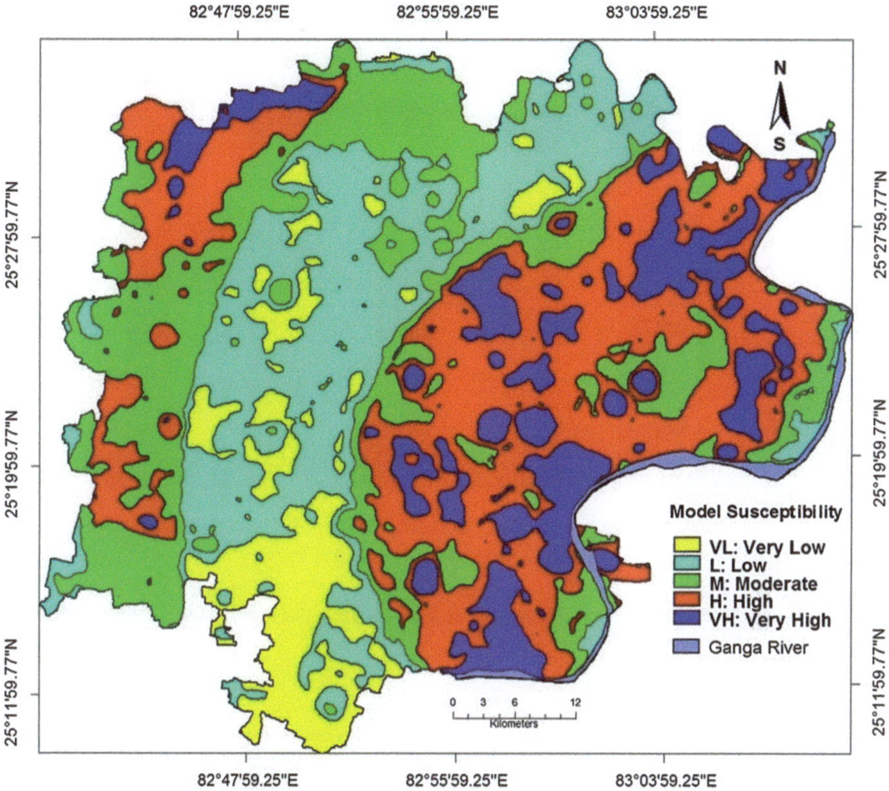

Fig. 6.25 Malaria Susceptibility Zone (MSZ) based on the information value method

distance to health facilities, distance to road, temperature, land cover and NDVI respectively.

The MSI values calculated from the weightage method are found in the range from 21 to 37 and (Fig. 6.28 and 6.29). It is cleared from the Fig. 6.28 that model index value 21–24 consist 1.71 % of the malaria pixels whereas 51.79 % and 14.35 % of the pixels comes in 27–30 and 33–37 model index classes respectively (Rai and Nathawat 2013). The cumulative frequency curve of MSI values has been classified into five classes representing near equal distribution to yield five malaria susceptibility zones i.e. very low, low, moderate, high and very high (Figs. 6.29 and 6.31).

In the Table 6.12 and Fig. 6.29, it is very well shown that 51.78 % of malaria area comes in moderate class where as 0.94 % and 3.98 % of the malaria area comes in very low and very high classes respectively (Fig. 6.30).

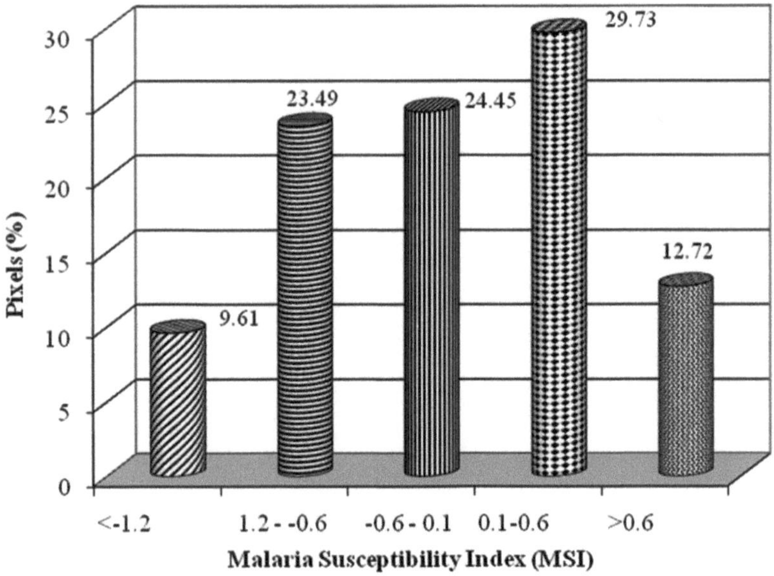

Fig. 6.26 Distribution frequency histogram of malaria based on the information value model

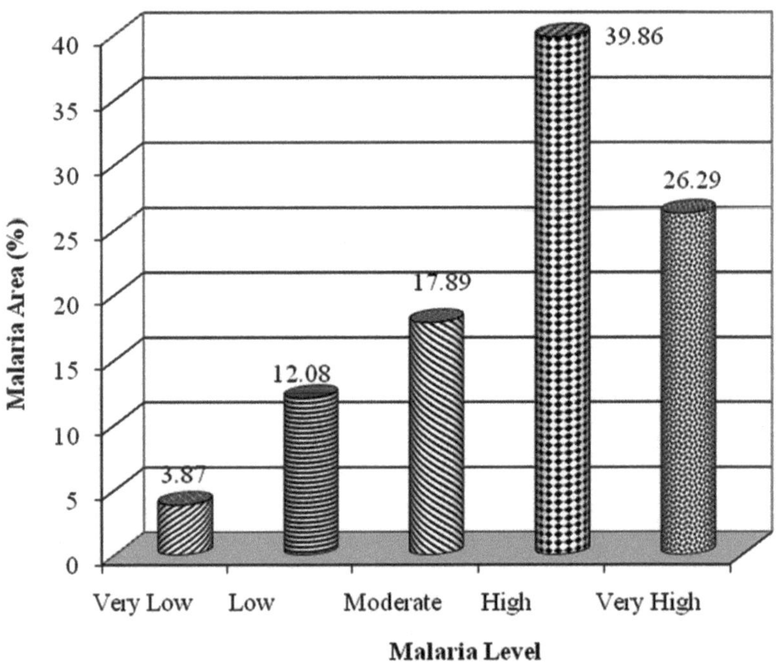

Fig. 6.27 Malaria area vs. malaria susceptibility based on the information value method

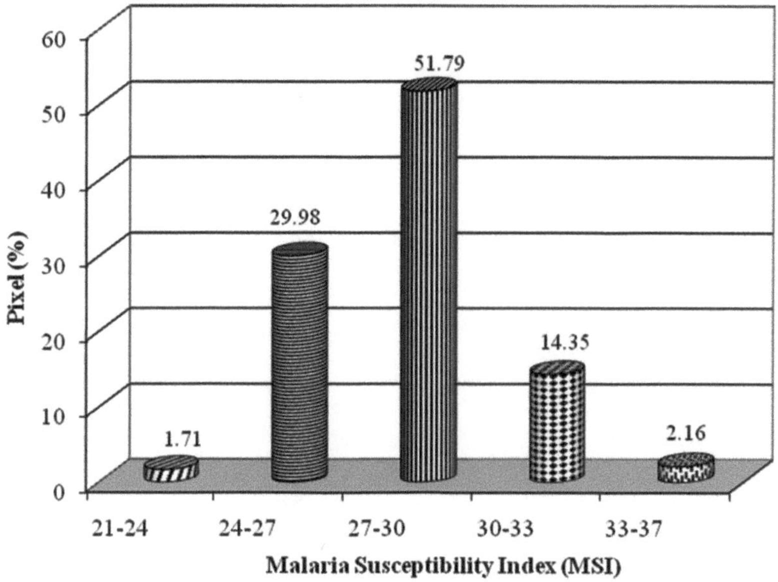

Fig. 6.28 Distribution frequency histogram of malaria based on the heuristic approach (Qualitative Map Combination)

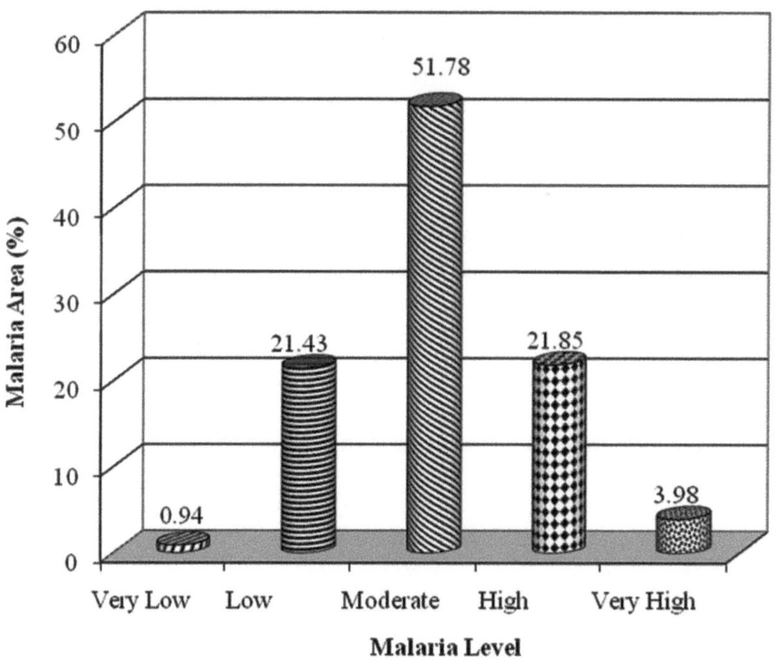

Fig. 6.29 Malaria area vs. malaria susceptibility level based on heuristic approach (Qualitative Map Combination)

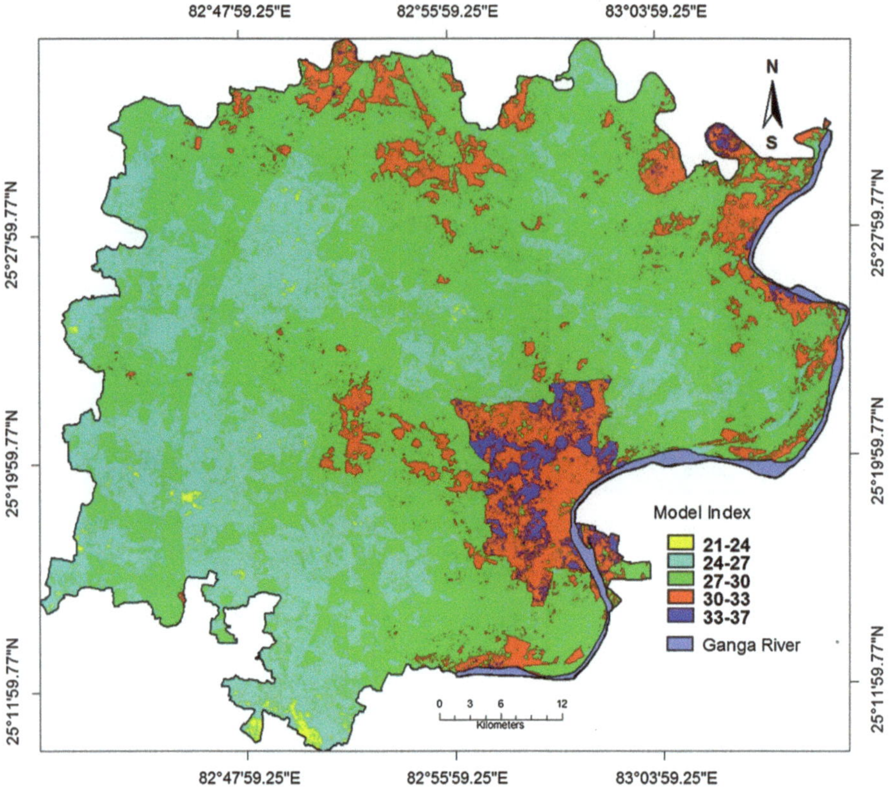

Fig. 6.30 Malaria Susceptibility Index (MSI) through heuristic approach (Qualitative Map Combination)

6.5 Comparison of Statistical Methods to Select Optimum Model for MSZ by Qs Method

The second aim of this study are choosing optimum model for malaria susceptibility zonation. For appraisal the results of the models, malaria density ratio (*Qs*) method has been used (Rai and Nathawat 2013):

To appraisal the results in this method (Eq. 6.7) has been used

$$D_r = \frac{d'}{d} = \frac{Sl'/Sl}{Sa'/Sa} \tag{6.7}$$

Where, D_r is the ratio of whole malaria, d' is the density of malaria in the susceptibility level, d is the malaria density in the area, Sl' is the malaria area pixels in special susceptible level, Sl is the area of special susceptible level, Sa' is the malaria

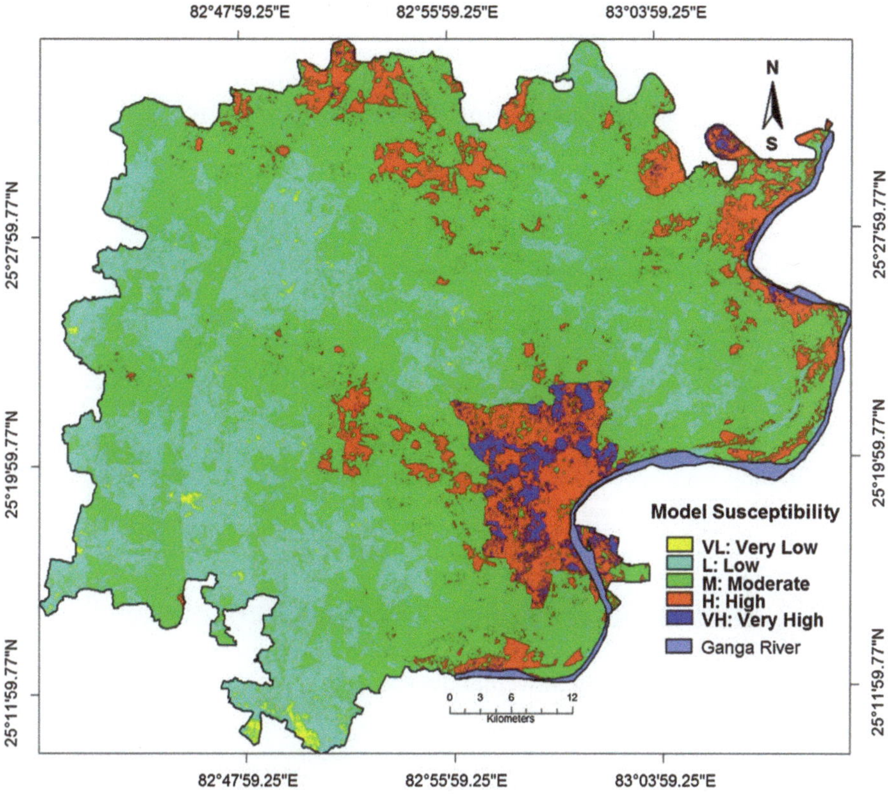

Fig. 6.31 Malaria Susceptibility Zone (MSZ) through heuristic approach

Table 6.12 Status of malaria area percentage vs. malaria level based on the heuristic approach (Qualitative Map Combination)

Malaria level	Total no. of pixels	Pixel (%)	Malaria area (sq.km)	Malaria area (%)
Very low	10,448	1.71	11.68	0.94
Low	183,170	29.98	265.46	21.43
Moderate	316,394	51.79	641.22	51.78
High	87,664	14.35	270.62	21.85
Very high	13,201	2.160	49.29	3.98
Total	611,390	100.00	1238.43	100.00

Table 6.13 Results of the Qs for the used models

Hazard level method	Very low	Low	Moderate	High	Very high	Qs
Information value	0.40	0.41	0.58	1.06	1.63	3.96
Qualitative map combination	0.55	0.57	0.79	1.20	1.46	1.67
Multiple linear regression	0.14	0.41	0.62	1.00	1.17	1.43

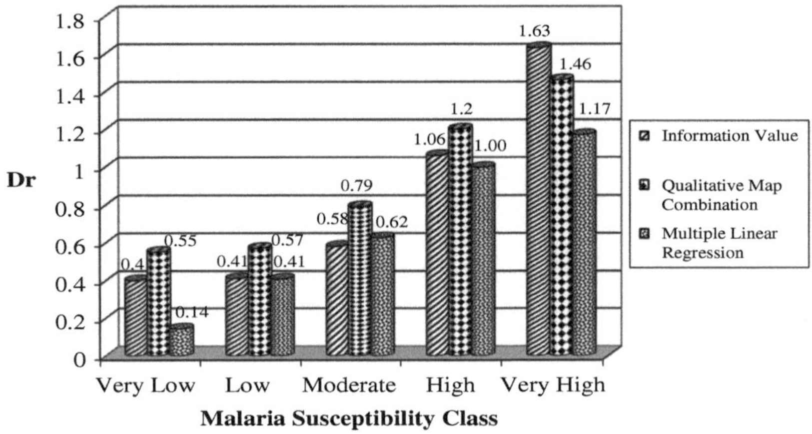

Fig. 6.32 Dr values of Qs method vs. malaria susceptibility class

area pixels in the study area and Sa is total pixels in the study area (Rai and Nathawat 2013).

$$Qs = \sum_{i}^{n} (D_r - 1)^2 \times \% S \tag{6.8}$$

Where, *S* is the ratio of the area of each susceptible level of the study area and *Qs* is the quality score of each models. Based on this equation (Eq. 6.8) the results have been clearly shown in the Table 6.13. Figure 6.32 shows that *Dr* Values for each models of malaria susceptibility class.

As given in the Table 6.13 and Fig. 6.33, it is calculated that the Information Value Method having *Qs* = 3.96 has been nominated as an optimal model for malaria susceptibility zonation, whereas *Qs* value for Qualitative Map Combination (Heuristics Method) and Multiple Linear Regression method are 1.67 and 1.43 respectively (Rai and Nathawat 2013).

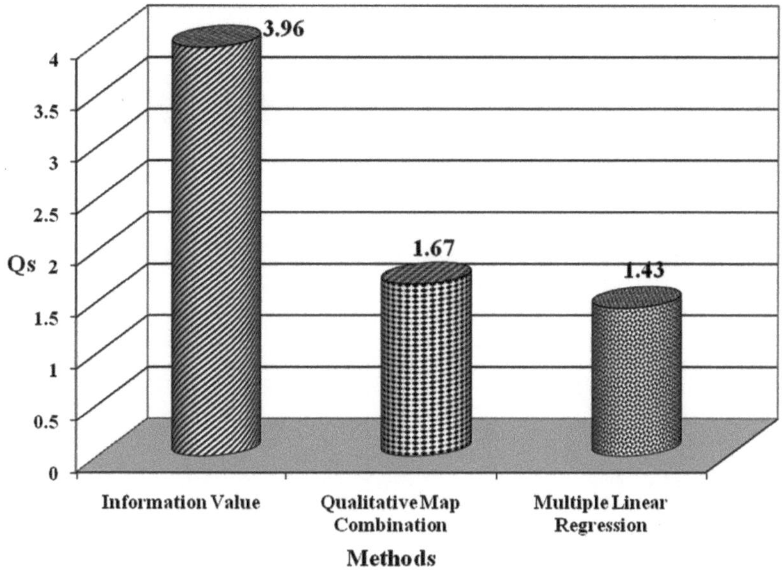

Fig. 6.33 Results of the Qs for the used models

6.6 Verification of the Susceptibility Methods by Area Under Curve (AUC)

The malaria investigation is completed using the information value, multiple linear regressions, and heuristic models, and the analysis outcomes are tested using the malaria locations for the study area; the outcome of this verification is shown in the Fig. 6.34 (Rai and Nathawat 2013).

The verification method is achieved by comparison of present malaria data and malaria analysis results. The comparison outcomes are shown in Table 6.14 and as a line graph in the Fig. 6.34 exemplify how well the estimators accomplish with respect to the malarias used in constructing those estimators (Rai and Nathawat 2013). Verification of the success rate is dependent on the malaria susceptibility analysis outcome using the malaria occurrence locations, for the three kinds of analysis methods—information value, multiple linear regression, and heuristic models. The rate curves are generated and the "areas under the curves" are calculated for all three cases of susceptibility maps using the existing malaria location data. The "areas under the curves" establishes one of the most regularly used accuracy statistics for the prophecy models in environmental assessments (Begueria 2006; Rai and Nathawat 2013). The rate clearly elucidates how well the model and factor envisage the malaria (Chung and Fabbri 1999; Rai and Nathawat 2013). So, the area under the curve can be used to evaluate the prediction accuracy qualitatively. To find the

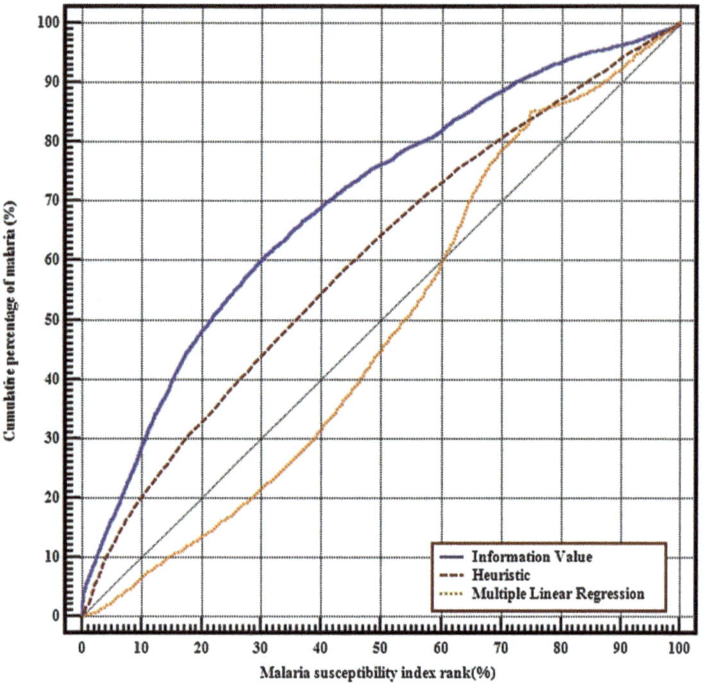

Fig. 6.34 Illustration of cumulative frequency diagram showing malaria susceptibility index rank (*x-axis*) occurring in cumulative percentage of malaria occurrence (*y-axis*)

Table 6.14 Verification and success rate for the study area

Range	Success rate curve (Information value)	Success rate curve (Heuristic)	Success rate curve (Multiple linear regression)
100–100	0	0	0
90–100	28	20	6
80–100	48	34	14
70–100	60	45	21
60–100	69	54	32
50–100	76	65	46
40–100	82	73	50
30–100	88	81	79
20–100	94	88	87
10–100	96	94	93
0–100	100	100	100

Table 6.15 Verification results using Area under Curve (AUC)

Method	AUC	Prediction accuracy (%)	Standard Error (SE)	95 % Confidence Interval (CI)
Information value	0.696	69.60	0.00357	0.691–0.701
Heuristic	0.603	60.30	0.00371	0.598–0.608
Multiple linear regression	0.484	48.40	0.00368	0.479–0.489

relative ranks for each prediction pattern, the calculated index values of all cells in the study area are organized in descending order. The ordered cell values are then divided into 100 classes and set on the y-axis, with accumulated 1 % intervals on the x-axis. The rate verification outcomes seem as a line in the Fig. 6.34 and Table 6.14 (Rai and Nathawat 2013).

In the case of the information value model used, 90–100 % (10 %) class of the study area where the malaria index had a higher rank could explicate about 28 % of all the malaria in the success rate and is designated as "very highly susceptible" zone (Rai and Nathawat 2013). The next 80–100 % (20 %) class of the study area where the malaria index had a higher rank could elucidate about 48 % of the malarias in the success rate and is classified as "high susceptible" zone. Similarly, the 60–100 % (40 %) class of the study area where the malaria index had a relatively lower rank could explain 69 % of the malaria in the success rate and is classified as "moderately susceptible" zone. Finally, the outstanding 40–100 % (60 %) class of the study area where the malaria index had a low rank could explicate 82 % of the malaria are classified as "not susceptible" zone. The same process is assumed for classification and verification of the hazard maps achieved through heuristic (Table 6.15, column 2) and multiple linear regression (Table 6.15, column 3) models. To relate the outcomes quantitatively, the areas under the curve are re-calculated as the total area is 1 which means perfect prediction accuracy. So, the area under a curve is used to assess the prediction accuracy qualitatively, as shown in the Fig. 6.34 (Rai and Nathawat 2013). From the Fig. 6.34, verification results show that in the information value case, the area under curve (AUC) is 0.696 and the prediction accuracy is 69.60 %. In the heuristic case, the AUC is 0.603 and the prediction accuracy is 60.30 %. In the multiple linear regression case, the AUC is 0.484 and the prediction accuracy is 48.40 %. So from the success rate graphs (the Fig. 6.34), it is relatively obvious that the information value has the greatest prediction accuracy of 69.60 %, while the multiple linear regression has the poorest accuracy of 48.40 %, with a difference of about 21.2 % (Rai and Nathawat 2013).

Even though, for the first seven classes (30–100 %), the heuristic model is better than from the multiple linear regression model, except for the remainder of the classes (70–100 %), the heuristic model produced rather similar outcomes to those from the information value and multiple linear regression models. The multiple linear regression model is inferior than the information value and heuristic models in all classes (Rai and Nathawat 2013).

References

Begueria, S. (2006). Validation and evaluation of predictive models in hazard assessment and risk management. *Natural Hazards, 37*, 315–329.

Chung, C. F., & Fabbri, A. G. (1999). Probabilistic prediction models for landslide hazard mapping. Photogrammetric Eng. *Remote Sensing, 65*(12), 1389–1399.

Colledge, M., Maxwell, H., Parker, R., & Morse, D. J. (1996). Geographic information systems (GIS) in general practice: a new tool for needs assessment. *Journal of Informatics in Primary Care*, 7–11.

Connor, S. J., Flasse, S. P., Perryman, A. H., & Thomson, M. C. (1997). *The contribution of satellite derived information to malaria stratification, monitoring and early warning*. World Health Organization mimeographed series. WHO/MAL/97.1079

Craig, M. H., Snow, R. W., & Le Sueur, D. (1999). A climate-based distribution model of malaria transmission in sub-Saharan Africa. *Parasitology Today, 15*, 105–111.

Davis, J. C. (1986). *Statistical and data analysis in geology* (2nd ed., p. 646). New York: Wiley.

Hoek, W., Konradson, F., Amersinghe, P. H., Perara, D., Piyaratne, M. K., & Amerasinghe, F. P. (2003). Towards a risk map of malaria in Srilanka: The importance of house location relative to vector breeding sites. *International Journal of Epidemiology, 32*, 280–285.

Irigaray, C., Ferna´ndez, T., El Hamdouni, R., & Chaco´n, J. (2007). Evaluation and validation of landslide-susceptibility maps obtained by a GIS matrix method: Examples from the Betic Cordillera (Southern Spain). *Natural Hazards, 41*, 61–79.

Kulldorff, M. (1998). Statistical methods for spatial epidemiology: Tests for randomness. In A. C. Gatrell & M. Löytönen (Eds.), *GIS and health* (pp. 49–62). London: Taylor and Francis.

Lawson, A. (2009). *Bayesian disease mapping. Hierarchical modeling in spatial epidemiology*. Boca Raton: CRC Press.

Omumbo, J., Ouma, J., Rapuoda, B., Craig, M. H., & Sueur, I. (1998). Mapping malaria transmission intensity using Geographical information Systems (GIS): An example from Kenya. *Annals of Tropical Medicine and Parasitology, 92*(1), 7–21.

Rai P. K., & Nathawat M. S. (2013). Application of GIS & statistical methods to select optimum model for malaria susceptibility zone and verification of the susceptibility methods by Area Under Curve (AUC): A case study, Scientific Annals of "Alexandru Ioan Cuza" University of Iasi – Geography series. *SciGeo Journal, Romania, 59*(2), 73–94. (ISSN: 1223–5334).

Rai, P. K., Nathawat, M. S., & Onagh, M. (2012). Application of multiple linear regression model through GIS and remote sensing for malaria mapping in Varanasi district, INDIA. *Health Science Journal (International Journal of Nursing Research and Review), 6*(4), 731–749. (ISSN 1108-7366E). Available at http://www.hsj.gr/volume6/issue4/6414

Rai P. K., Nathawat M. S., & Rai S. (2013). Using the information value method in Geographic Information System (GIS) and remote sensing for malaria mapping; a case study from India. *Informatics in Primary Care (University of Surrey, U.K.), 21*(1), 43–52. (ISSN 1475–9985). Available at http://dx.doi.org/10.14236/jhi.v21i1.38

Riedel, N., Vounatsou, P., Miller, J. M., Gosoniu, L., Chizema-Kawesha, E., Mukonka, V., & Steketee, R. W. (2010). Geographical patterns and predictors of malaria risk in Zambia: Bayesian geostatistical modeling of the 2006 Zambia National Malaria Indicator Survey (ZMIS). *Malaria Journal, 9*, 37.

Roper, M. H., Torres, R. S. C., Goicochea, C. G., Anderson, E. M., Guarda, J. S. A., Calampa, C., Hightower, A. W., & Magill, A. J. (2000). The Epidemiology of malaria in epidemic area of the Peruvian Amazon. *American Journal Tropical Medical Hygiene, 62*(2), 247–256.

Shah, N. S. (2007). Burning quest: Should public health be sacrificed to waste? *Down To Earth*, 25–26.

Sudhakar, S., Srinivas, T., Palit, A., Kar, S. K., & Battacharya, S. K. (2006). Mapping of risk prone areas of kala-azar (*Visceral leishmaniasis*) in parts of Bihar state, India: an RS and GIS approach. *Journal of Vector Borne Disease, September, 43*, 115–122.

Sweeney, A. W. (1997). A spatial analysis of mosquito distribution. *GIS User, 21*, 20–21.

Van Westen, C. J. (1993). *Application of geographic information system to landslide hazard zonation*. ITC-Publication No. 15, ITC, Enschede, p. 245.

Chapter 7
An Analysis of Geographical Survey for Utilization of Health Care Facilities

Abstract Utilization of health services is a complex phenomenon which, on the hand, is influenced by the awareness by an individual of the need for services thereby endorsing him to take a choice to use them and, on the other hand, by the availability, accessibility and organizational characteristics of health care services itself. The main objective of this chapter is to estimate the utilization pattern of health care services in the Varanasi district of India. Primary data pertaining to the utilization of health care facilities are collected from 800 respondents of 16 selected villages of rural Varanasi and analyzed with the SPSS statistical software. Varanasi City proper was not considered for this purpose because the presence and functioning of many private and government hospitals here meant that people were able to avail themselves of a fairly good range of healthcare facilities in comparison to people residing in the rural areas. Results of the findings revealed a high level of awareness among the local public of both the existence of the health care centres (78 %) and the type of health services they provided (75 % for vaccination; 70 % mother-child health (MCH) services; 62 % family planning; and 52 % general treatment). Despite such high levels of awareness only 25 % of them are satisfied with all the health care services provided by the primary health centres (PHC), 60 % are only partially satisfied and the remaining 14 % were not satisfied at all. These findings thus underline the geographical disparities in health facilities between urban and rural area of Varanasi.

Keywords Health services • Varanasi • Primary health centres (PHC) • Primary data • SPSS • Mother-child health (MCH) • Family planning

7.1 Utilization of Health Care Service

Level of utilization pattern of health care facilities obviously designates the consciousness and attitude of people towards their health (Prakasam 1995). Education, economy, sex and social status are main swaying factors for utilization of health care services. An educated person is more vigilant about his health than an uneducated. Females use these services less as compared to males (Sinha and Rajeswari 1993; Rai and Nathawat 2014).

P.K. Rai, M.S. Nathawat, *Geoinformatics in Health Facility Analysis*,
DOI 10.1007/978-3-319-44624-0_7

Utilization of health services is a complex phenomenon which, on the hand, is influenced by the awareness by an individual of the need for services thereby endorsing him to take a choice to use them and, on the other hand, by the availability, accessibility and organizational characteristics of health care services itself (Murali 1981, Sinha and Rajeswari 1993; Rai and Nathawat 2014). Besides, effective utilization of health care services depends on dependability, awareness, motivation and finally on the perception of the people about the facilities and the need about a specific service. Failing to which, by passing phenomenon may take place (Aday and Andersen 1974; Srinivasan 1984; Kumra and Singh 1994; Rai and Nathawat 2014).

In rural areas the health care services and family welfare programmes are being provided through the network of primary health centres (PHC's) and sub-centres (Datta and Kale 1969; Rai and Nathawat 2014). If one looks at the present availability heart centres, heart breaths with satisfaction. Despite, the country has not been able to attain the target goal of 'health for all'. For reasonable explanation, the utilization of health facilities is obligatory to be investigated in details (Mohler 1978; Ghosh and Mukharjee 1989; Rai and Nathawat 2014).

The government offers curative, preventive and primitive health services facilities through PHC's and sub-centres. PHC's provide a variety of services while sub-centres only family welfare services and some primary treatment (McPhail et al. 1963). The current study displays that immunization and vaccination facilities are more widespread in public community (Akhtar and Khan 1993; Rai and Nathawat 2014).

To bring out the pattern of health care utilization in the study area, in this chapter an effort has been made to firstly the utilization of health care facilities on the basis of government archives and secondly it has been determined through views of 800 surveyed respondents (Rai and Nathawat 2014).

7.2 Vaccination and Immunization

It is used for the growth of immunity power in human body. More than 20 different immunizations are not obtainable in the world. But all of the immunization is not available in India (Basu 1984). Immunization may be divided into two groups:

1. Primary Immunization: For the development of immunity power the body, one or more than one times vaccine is injected. After a certain interval immunity power becomes less then comparatively better immunization is required (Rai and Nathawat 2014).
2. Secondary Immunization or Boosters: It is an extra dose of immunization. Normally many of the people who are acquainted with primary immunization forget to take secondary immunization. But it is duty of the concerned health workers to say the immunized people about the booster's dose at definite intervals (Rai and Nathawat 2014).

7.2.1 *Infant Immunization*

Infant are more susceptible to diseases. So for the prevention of diseases and afterword's medical technology has invented many types of vaccines/solutions. Immunization may be divided into two parts. Vaccines immunoglobulis-BCG, typhoid, plague, polio, measles, influenza, diphtheria, tetanus, rabiz, etc. such diseases may be prevented through immunizing infants. In rural areas people don't pay the obligatory care towards the mother and child health care (Rai and Nathawat 2014). Pre and post precautions during and after the delivery is not acceptable in rural areas due to ignorance, illiteracy and unaffordability. On account of poor information among people, depression of health workers and often unavailability of vaccines many children don't get vaccine at right time. Under the prevailing situation it becomes the responsibility of PHC's and sub-centres to take all sorts to save the lives of mother and child (Khan 1980; Rai and Nathawat 2014).

7.2.2 *Extensive Vaccination and Immunization Programme*

Following vaccination and immunization programme has been launched by the government to save the lives of children and pregnant women:

7.2.2.1 DPT (Triple Antigen)

It is preventive vaccine for diphtheria, kali khasi and tetanus. Normally, it is given to 3–9 months old child. In primary prevention, after 4–6 weeks of birth three injections are vaccinated (Rai and Nathawat 2014). After few months their effects, becomes lesser, then third primary injection to 1 year old child is vaccinated. After that every third year injection vaccine should be taken (Rai and Nathawat 2014).

7.2.2.2 ATS (Anti Tetanus Syrup)

For injured person tetanus toxide and DPT doses are obligatory.

7.2.2.3 Polio

The vaccine of polio had come into effect during 1978–1980. So far there has not developed any medicine for polio. Therefore, vaccination of polio is must. It starts at the age of 2–3 months of child. Three doses of polio are given at an interval of 4–8 weeks.

7.2.2.4 BCG

It is used as the time of birth after the test of tuberculin. When tuberculin test is found negative, BCG can be given to child.

7.2.2.5 Measles

It is specified from 1985. After 9 months to 2 years old children it can be given one time.

7.2.2.6 Typhoid

It is given to children after 2 years their birth. Generally this vaccine is injected two or three in a child at 1 or 4 week interval.

7.2.2.7 Cholera

It is given after 1 or 2 years of birth of child. It is suitable in summer season. In the beginning 2 or 3 doses are injected in a child at 1–4 weeks interval (Rai and Nathawat 2014).

7.2.2.8 Extensive Immunization from 2005–2006 to 2009–2010

Table 7.1 shows that surviving children different preventive does of vaccines at certain intervals for various purpose. Government is also paying due attention towards this direction. To substantiate, the data was collected for four consecutive years i.e. form 2005–2006 to 2009–2010 (Table 7.2). The data show differential figures of given doses. In 2005–2006 the target was set to immunize 51,204 children and achievement was 105.81 %, 100.09 % and 100.59 % for DPT and OPV I, II and III doses respectively. The achievements for BCG and measles were 101.59 % and 91.4 % respectively. During 2005–2006 to 2009–2010 achievement has always been higher than the target (Rai and Nathawat 2014). Only immunization of meseals show lesser achievement than the set target. But it also shows an increasing trend. These statistics reflect an increasing awareness of health care for their children in people of the study- area (Table 7.2).

7.2.2.9 Pulse Polio Immunization

This programme has gained wide popularity both in rural and urban areas. It is given to children of below 5 years of age. Aim of this programme is to get total control on polio. In winter season one dose of oral drop of polio is given to all the

Table 7.1 National immunization programme

User	Age	Vaccine	Dose	Disease
Infant	3–9 months	D.P.T.	3 doses	Diphtheria, Kali Khasi, Tetanus
		Polio	3 doses	Polio
		B.C.G.	3 doses	T.B.
	9–12 months	Measles	1 dose	Measles
	18–24 months	D.P.T.	1 dose	
		Polio	1 dose	
Child	5–6 years	D.T.	1 dose	
	10 years	T.T.	1 dose	Tetanus
		Typhoid	2 dose	Typhoid
	16 years	T.T.	1 dose	
		Typhoid	1 dose	
Pregnant women	16–36 weeks	T.T.	1 dose (If two dose are not injected)	

Source: Jansankhya Shiksha, Sidhanta Avan Tatwa, Chapter Family Welfare Programme, Jansankhya Kendra, U.P., Lucknow, U.P., India

children below 5 years of age in the whole country. Many working booths are created for this purpose (Rai and Nathawat 2014). It is remarkable to mention that target was fixed to cover 3,77,506 children under polio immunization programme in Varanasi district for the year 2009–2010. But I and II round immunization programme achievement data of 2009–2010 were higher than the target in majority of the development blocks of the district (Table 7.3). The achievement of only two blocks, namely Chiraigaon (93.89 %) and Harhua (93.78 %) was less than the target set for I round. In IIIrd round many blocks of the district could not achieve their targets. Only two blocks namely Cholapur (102.34 %) and Araziline (100.19 %) could programme were recorded as 100.26 %, 102.87 % and 99.70 % for I, II, III round respectively. For this purpose, altogether 977 teams were arranged at reasonable locations in the district (Rai and Nathawat 2014).

7.3 Utilization of Family Welfare Programme

Family planning programme was initiated as early as in 1951 to control always increasing population of the country. In 1970s family planning drive was started more strictly and effectively. During this period some sort of fear related to bad effects and impotency was spread in the society and community (Rai and Nathawat 2014).

But now this programme is running fear and so has wide acceptance. During Seventh Five Plan (1984–1989) this programme was changed into family welfare programme. During Ninth Five Year Plan much emphasis was given on the mother and child health care (MCH Programme).

Table 7.2 Immunization status in Varanasi district from 2005–2006 to 2009–2010

	OPV (Children 0–1 year)					DPT (Children 0–1 year)					BCG		Meseals	
Year	Target	I	II	III	Booster	Target	I	II	III	Booster	Target	Achiev	Target	Achiev
2006–2007	51204	54183	51255	51508	20266	51204	54183	51249	51498	20266	51204	52020	51204	46808
		105.8	100	100.59			105.81	100.08	100.57			101.59		91.41
2007–2008	52125	55149	52768	52402	21765	52125	54659	52715	52366	21676	52124	52666	52125	48358
		105.8	101.32	100.53			104.86	101.13	100.46			101.03		92.77
2008–2009	51640	59050	55607	55128	19596	51640	58954	55622	54979	19485	51640	57114	51640	51152
		114.34	107.68	106.75			114.16	107.71	106.46			110.6		99.05
2009–2010	49793	54977	52577	53078	21893	49793	54977	52577	53078	21893	49793	5578	49793	49778
		110.4	105.59	106.59			110.4	105.59	106.46			111.61		99.96

Source: CMO office, Varanasi district, 2009

Table 7.3 Pulse polio immunization for below 5 years children's, 2008–2009

Development block	I Round				II Round				III Round			
	Target	Achiev	% Achiev	No. of team	Target	Achiev	% Achiev	No. of team	Target	Achiev	% Achiev	No. of team
Baranagaon	42488	44996	105.9	111	44996	45239	100.54	111	45239	45087	99.99	111
Pindra	405144	40686	100.42	106	40686	41142	101.12	106	41142	40563	98.59	106
Cholapur	40459	40465	100.01	106	40465	42232	104.36	106	42232	43222	102.34	106
Chiraigaon	53455	50191	93.86	140	50191	52360	104.32	140	52360	52325	99.93	140
Harhua	44844	42035	92.73	105	42035	45707	108.73	105	45707	45232	98.96	105
Sewapuri	41506	41584	100.18	109	41584	41628	100.1	109	41628	41620	99.98	109
Araziline	65553	66474	101.4	172	66474	67425	101.1	172	67425	57557	100.19	172
Kashi Vidyapith	48687	52085	106.97	128	52085	53648	103	128	53648	52608	98.06	128
Total	377506	378516	100.26	277	378516	389381	102.87	977	689381	388214	99.7	977

Source: CMO office, Varanasi district, 2009

7.3.1 *Immunization to Pregnant Women*

PHC's and sub-centres provide immunization facilities for pregnant women. At the time of registration of pregnancy first dose of T.T and after 1 month second dose of T.T. are given to pregnant women. If due to any cause the patient does not get first and second dose of T.T., then booster dosage is given (Rai and Nathawat 2014). Table 7.4 shows the beneficiaries of T.T. immunization for the period of 2005–2009 in Varanasi district.

7.3.2 *Delivery*

PHC's and sub-centre do provide delivery facilities. In addition, Auxiliary Nurse Midwife (ANM) and trained dai (mainly women) make delivery at home. But medical expert favour institutional delivery so as to control the problem arised out at the time of delivery. Table 7.4 shows that maximum delivery (52.84 %) was carried out by ANM while 22.34 deliveries were performed at PHC and their sub-centres. In 2007–2008 about 36,171 cases of delivery were reported in which the contribution of ANM was 52.04 % followed by 24.61 % institutional delivery, 23.24 % by trained dai and 0.09 % by other sources but in 2008–2009 only about 34,361 cases of delivery were recorded (Rai and Nathawat 2014).

Table 7.4 TT immunizations in pregnant women and delivery

	TT immunization in pregnant women			Delivery type					
Year	I	II	Busters	Institutional delivery	Serious delivery	ANM	Trained Dai	Other	Total
2004–2005	43305	42658	1467	7631	457	18609	6368	200	33265
				(22.94)	(1.37)	(55.94)	(19.14)	(0.6)	(100.00)
2005–2006	44324	42436	2129	11225	–	–	–	–	–
2006–2007	49568	48538	1913	9624	803	20925	7560	1758	40670
				(23.66)	(1.97)	(51.45)	(18.58)	(4.32)	(100.00)
2007–2008	50457	47922	1128	8167	735	18825	8408	36	36171
				(22.57)	(2.03)	(52.04)	(23.24)		(100.00)
2008–2009	50115	48458	1037	6944	585	17846	8986		34361
				(20.2)	(1.7)	(51.93)	(26.15)		(100.00)

Source: CMO office, Varanasi district, 2009

7.3.3 Distribution of Vitamin A and Iron Tablets

PHC's also deliver Vitamin A and Iron tablets to the villagers living in their territory to control occurrence of night blindness and anaemia respectively. Table 7.5 displays year-wise distribution of iron and vitamin A tablets in rural areas of Varanasi district. During 2005–2006, 23,498 persons were given iron large tablets in first stage which increase to 48,277 persons during 2008–2009 periods. Similarly iron small tablets were distributed to 31,694 to persons during 2005–2006, which increase to 48,840 persons in 2008–2009. Data regarding distribution of Vitamin A tablets that a significant number of villagers were given Vitamin tablets to control night blindness.

Data regarding block-wise distribution of iron and Vitamin A tablets were also collected at block level (Table 7.6). Maximum numbers of villagers having given iron large tablets in first stage were found in Kashi Vidyapith block (7197 persons) whereas minimum in Sewapuri block (3966 persons) during 2009 period. Further maximum number of person (7581) who obtained in Vitamin A tablets form PHC's were recorded in Araziline block while minimum (1902) in Baragaon block in first stage. The distributional pattern of Vitamin A and Iron tablets at block level reveals the efficiency of PHC's in mitigating deficiency diseases in the study-area.

7.3.4 Contraception

Government provides various contraception facilities such as sterilization, IUD, oral pills and condom through PHC's.

7.3.5 Treatment

It is very well clear that immunization and contraception facilities rendered by PHC's and sub-centres are more common in the district but clinical services are not good. People are not pleased with available clinical services. Now a day's medical technologies have developed a lot but PHC's are running with traditional facilities. They are ill equipped terms of instruments, medicines and diagnostic technologies. There has been made a provision of two doctors for each PHC. Only block level PHC's have blood test facilities and that too only for malaria parasite and sputum for AFB. There is no x-ray facility. In contrary majority of private hospitals are endowed with all sorts of Blood Test, X-ray, Ultrasonography, ECG, Echocardiogram and CT scanning etc. Before the start of treatment, doctors as well as patients both needs the through checkup of the problem. As such the PHC's should be equipped with maximum possible labs for testing.

Table 7.5 Year wise distribution of iron and vitamin a tablets (2005–2009)

Year	Iron large			Iron small			Vitamin A				
	I	II	II	I	II	III	I	II	III	IV	V
2005–2006	23498	20998	20014	31694	27924	27893	49653	17304	9825	4570	13473
2006–2007	25533	23996	26084	33670	30419	28447	39481	19649	10775	6399	5778
2007–2008	42845	42501	42329	40331	28892	30452	52321	23885	11343	7606	5864
2008–2009	48277	48840	50015	40049	33336	30629	40336	19406	11957	6467	6424

Source: CMO office, Varanasi district, 2009

Table 7.6 Block-wise distribution of iron and vitamin tablets (2009)

Development block	Iron large			Iron small			Vitamin A				
	I	II	II	I	II	III	I	II	III	IV	V
Baranagaon	5006	5006	5006	3745	3745	3745	1902	581	208	36	408
Pindra	6787	6787	6787	4255	4255	4255	2394	–	–	–	–
Cholapur	5687	5687	5687	5983	5983	5983	5025	2982	1149	496	308
Chiraigaon	6507	5366	4899	8131	4610	3992	6414	5302	4365	1836	1449
Harhua	6046	6201	6201	4610	2864	2372	5981	2656	254	–	–
Sewapuri	3966	6375	8795	–	–	–	6545	2408	1913	1757	1714
Araziline	7081	6528	6136	7281	6599	5761	7581	4053	3504	2251	2453
Kashi Vidyapith	7197	6890	6503	6047	5280	4521	4494	1424	564	91	92
Total	48277	48840	50014	40049	33336	30629	40336	19406	11957	6467	6224

Source: CMO office, Varanasi district, 2009

Table 7.7 displays the monthly treatment of patients at different PHC's in 2009. In 2009, the highest number of patients received treatment at Cholapur PHC (12,577 persons). There is not found any consistency in turn out of outdoor patients in different seasons/months of the year at block level (Table 7.7). However, on the basis of turnout of patients, utilization level has been assessed in the following paragraphs.

(i) Poor Utilization: It is very fascinating to note that the utilization of PHC's for treatment is poor at those PHC's which are lying either in the vicinity of Varanasi city or urban areas. For example in the said year the lower utilization of PHC's facilities for treatment is found at PHC of Baragaon (2703 patient), Kashi Vidyapith (3280 patients), Harhua (3851 patients) and Chiraigaon (6469 patients). The motives attributed to the poor utilization of PHC's facilities for clinical purposes are accessibility of better facilities at private hospitals at reasonable and rates (Rai and Nathawat 2014).

(ii) Better Utilization: The PHC's located in distant areas have registered more crowds of patients. For instance, Pindra (9279 patients), Sewapuri (8869 patients) and Araziline (8997 patients) PHC's shows better utilization of health care services. In many blocks there are no good private hospitals. So, the people have to go at PHC's/CHC's of their area. Secondly, poor income conditions of villagers do not allow avail facilities at city area (Rai and Nathawat 2014).

From temporal point of view data show that the maximum utilization of health care services has been obtained in 3 months namely July, August and September. These months belong to rainy season and characterized by vertical scorching sunlight, adequate rainfall and high humidity which give birth to malaria, filarial, typhoid and bacterial and viral infections. During these months accessibility too becomes poor. December, January and February record minimum number of patients due to better weather conditions from disease point of view. If the patient is serious at PHC's or CHC's, doctors refer him to the district hospitals, Varanasi. Specialized facilities with specialist doctors are available over there along with x-rays, pathology, blood bank and gynaecological facilities. In addition, there is separate TB hospitals and TB control unit equipped with medicine, test facilities and indoor care. Leprosy control unit provides services through PHC's, New PHC's, CHC's, sub-centres and district hospitals (Rai and Nathawat 2014).

7.4 The Utilization of Health Care Facilities in Opinion of Respondents Through Personal Survey

The simple matter regarding the utilization care services is to evaluate the popularity of primary health centres for providing desirable facilities to the nearby population (Mishra 1988). From this point of view, opinion of arbitrarily selected 800 respondents representing all the blocks has been used and their outcomes are

Table 7.7 PHC's wise OPD patients in each block in different months in 2009

Blocks	January	February	March	April	May	June	July	August	September	October	November	December	Total
Baragaon	120	145	278	334	334	252	389	491	104	105	53	98	2703
Pindra	520	413	567	550	590	840	1267	1677	960	982	562	351	9279
Cholapur	760	805	1030	998	1667	1112	1566	1543	1244	1133	431	288	12577
Chirai Gaon	340	268	338	445	445	621	876	811	895	700	445	285	6469
Harhua	224	234	229	334	334	413	432	434	350	341	280	246	3851
Sewapuri	433	567	778	667	776	988	1098	987	815	744	504	512	8869
Araziline	670	689	889	889	552	555	1036	1281	973	744	431	288	8997
Kashi Vidhyapith	210	186	377	223	226	414	323	349	268	253	230	221	3280

Source: CMO office, Varanasi district, 2009

studied with the support of SPSS statistical software (Fig. 7.1). It is clear from the study that immunization and vaccination services are more widespread than other services (Rai and Nathawat 2014).

The utilization arrangement of health care services designates that the people of the study-area utilized vaccination and immunization in considerable proportion (74 %). Next to this in popularity and tradition stands mother and child health care followed by treatment and family planning facilities. Jakhini village (Araziline block) owed to high approachability from Araziline PHC's as well as community centre come on top position with regard to utilization of vaccination and immunization facilities (Table 7.8). MCH facility has also been utilized by the respondents of Jakhini village of Varanasi (Rai and Nathawat 2014). Highest number of respondents of Rampur village has go to Cholapur CHC for the treatment of their disease while the utilization of family welfare facilities is identified maximum in Purai Kala-Harhua Village of Harhua block PHC's (Fig. 7.1).

The utilization of pathological lab facilities is very poor. Out of the 800 respondents, only 14 respondents (1.75 %) have used this facility (Table 6.8). Likewise, the diagnostic services have least fascinated the villagers. Now a day's people want to become assure about the illness before treatment. Only block level PHC's have such type of test facilities. Additional PHC's are deprived from such type of services. Test facilities include only blood smear test, malaria parasite and sputum test for

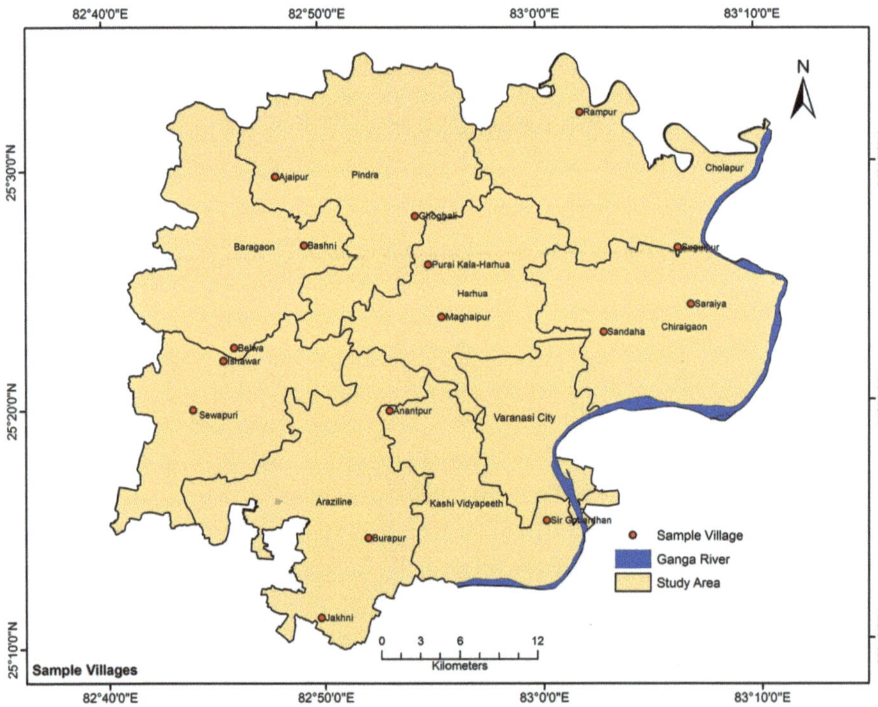

Fig. 7.1 GPS location of sampled villages selected for survey in rural part of Varanasi district

Table 7.8 Respondents opinion regarding utilization of health care facilities in selected villages in each development block

Development blocks	Villages	Treatments	MCH	Family planning	Disease control	Vaccination	Laboratory	Total respondents
Baragaon	Bashni	26	26	20	5	39	2	50
	Belwa	23	25	22	6	42	–	50
Pindra	Ajaipur	23	24	23	10	45	–	50
	Goghri	20	23	21	7	39	–	50
Cholapur	Sugulpur	18	20	18	5	36	–	50
	Rampur	36	31	25	15	28	8	50
Chirai Gaon	Sandaha	30	28	30	12	37	–	50
	Bankat	16	20	17	6	26	–	50
Harhua	Purai Kala Harhua	31	31	35	15	43	–	50
	Maghaipur	17	19	20	4	41	–	50
Sewapuri	Sakalpur	16	19	20	7	39	–	50
	Ishawar	25	27	25	10	40	–	50
Araziline	Jakhini	31	35	27	11	46	4	50
	Burapur	19	20	19	8	30	–	50
Kashi Vidhyapith	Anantpur	29	29	24	10	34	–	50
	Sir Gobardhan	20	22	18	5	27	–	50
Total		380 (47.5 %)	399 (50 %)	364 (45.5 %)	136 (17 %)	592 (74 %)	14 (1.75 %)	800

Source: Based on personal survey & self computed, 2010

AFB. Pathological test, x-ray and USG services are not obtainable on PHC's. However, X-ray facilities are obtainable at three CHC's (Cholapur, Araziline and Birawankot) of the area but looking the size and population of the study-area it is not adequate. Only 17 % respondents have visited their respective PHC's or New PHC's/CHC's for illness control (Rai and Nathawat 2014).

7.4.1 Distance Wise Respondent's Opinion Regarding Health Care Facilities

7.4.1.1 Distance Wise Respondent's Opinion Regarding Utilization of Health Care Facilities

It has been described by the academician and administrators that the distance and approachability affect the magnitude and frequency of utilization of health care facilities. As such it develops an important factor to examine the health care services (Rai and Nathawat 2014). For this purpose samples have been derived from the villages lying within 1 km, 1–3 km, 3–5 km and more than 5 km distances (Fig. 7.2).

Table 7.9 demonstrates distance-wise respondent's opinion regarding utilization of health care services. It discloses that maximum utilization of health care facilities are seen in the case of those villages which lie within less than 1 km. distance from PHC's.

About 99 % respondents of these villages have utilized vaccination facilities, 95 % for MCH, 80.5 % for treatment, 72 % for family planning and 35 % for disease control (Table 7.9). In contrast the minimum planning utilization of health care services is identified in the case of those villages which lie at a distance of more than

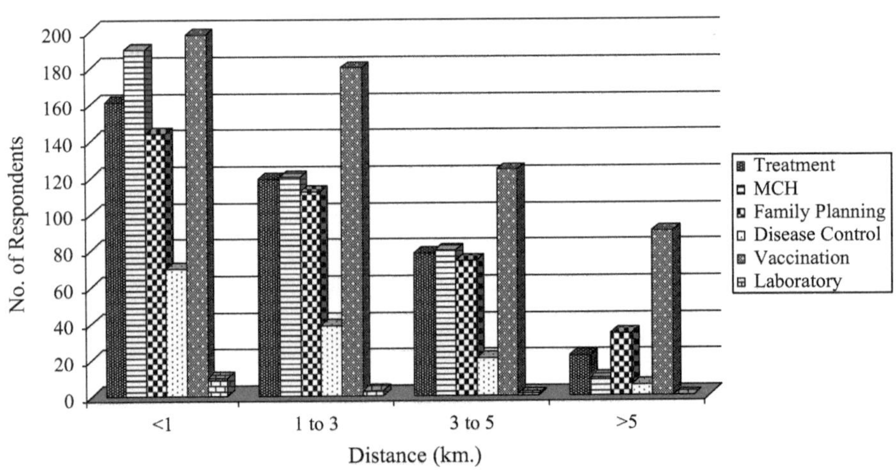

Fig. 7.2 Distance-wise respondents opinion in rural area of Varanasi district regarding utilization of health care facilities, 2010

Table 7.9 Distance wise respondents opinion regarding utilization of health care facilities

Distance (km)	Treatment	MCH	Family planning	Disease control	Vaccination	Laboratory	Total respondents
<1	161	190	144	70	198	10	200
1–3	119	120	112	39	180	3	200
3–5	78	80	74	21	124	1	200
>5	22	10	34	6	90	–	200
Total	380	400	364	136	592	14	800

Source: Based on personnel survey, 2010

5 km from their respective PHC's. It is also very well seen that about 45 % respondents have used vaccination facilities, 5 % respondents are taking facility for MCH, 11 % visited for treatment, 13 % go for family planning and 3 % for disease or illness control (Rai and Nathawat 2014).

7.4.1.2 Distance and Treatment Facilities

It is very clear from the study that distance decide the frequency of visits for utilising various health care services available at PHC's/CHS's. Table 7.10 demonstrates that 42.36 % of the total surveyed respondents come from long distances (>5 km) could have made their visit for availing treatment at PHC's. The respondents belonging to 1–3 km range shows 31.31 % utilization while the respondents of 3–5 km range displays 31.31 % utilization while the respondents of 3–5 km range marks 21.4 % utilization. It shows that with every increase in distance, there will be decreasing rate of utilization of health care services (Fig. 7.3). The treatment at PHC's comprises treatment for fever, diarrhoea, injury and other seasonal glitches. In all the cases of treatment the number of patients goes on decreasing with an increase in distance. Non availability of specialized doctors and poor diagnostic services also pay much in taking decision, particularly respondents come from long distance (Rai and Nathawat 2014).

7.4.1.3 Distance and Mother Child Health (MCH) Facilities

Distance wise respondents opinion regarding utilization of MCH facilities is given in Table 7.11. Out of 800 surveyed respondents, 50 % have utilized this facility. Among these, 47.5 % respondents were from within 1 km, 30 % from 1 to 3 km, 20 % from 3 to 5 km and 2.5 % users had come after travelling more than 5 km distance from PHC's and sub-centres (Rai and Nathawat 2014). Pre natal first time cases for medical supervision and injection were identified more (49.50 %) from less than 1 km distances whereas 37.12 % such respondents had come from 1.3 km range, 9.9 % were from 3 to 5 km and only 3.46 % cases were found from more than 5 km distance from PHC's (Fig. 7.4).

Table 7.10 Distance wise respondents opinion regarding utilization of treatment facilities

Distance (km)	Total treated	Fever	Diarrhoea	Injury	Other	Untreated	Total respondents
<1	161	60	60	34	41	25	200
1–3	119	22	39	17	28	95	200
3–5	78	18	15	9	18	112	200
>5	22	5	6	5	3	178	200
Total	380	105	120	65	90	420	800

Source: Based on personnel survey, 2010

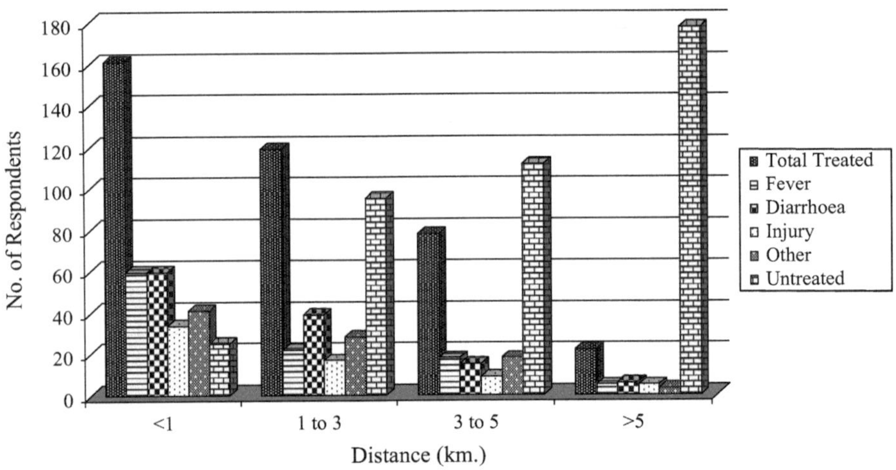

Fig. 7.3 Distance-wise respondents opinion in rural area of Varanasi district regarding utilization of treatment facilities, 2010

Pre-natal second times cases are noted less than Pre-natal first time cases. Out of 400 users, only 133 cases were distinguished for the pre-natal second time medical checkup. In this 48.87 % cases were documented from less than 1 km distance from PHC's/sub-centres, 36.87 % from 1 to 3 km, 10.52 % from 3.5 km and 3.75 % respond ants travelled more than 5 km distance respectively. Further, out of 400 medical advice seekers 145 cases of delivery were done at PHC's and sub-centres. In this 53.10 % delivery cases had travelled less than 1 km distance, 27.58 % 1–3 km, 16.55 % 3–5 km and 2.75 % respondents moved more than 5 km distance from PHC's and sub-centres (Rai and Nathawat 2014).

Post-natal first time help data display that maximum utilization has been completed by the respondents belonging to less than 1 km range from PHC's. In all 163 recorded cases for post-natal first time benefit, 50.92 % had come from within 1 km distance, 31.905 from 1 to 3 km, 12.26 % from 3 to 5 km and 4.90 % from more than 5 km distance from their own PHC's/sub-centres (Rai and Nathawat 2014).

Table 7.11 Distance wise respondents opinion regarding utilization of MCH facilities

Distance (km)	Total no. of MCH cases	Pre-natal first time	Pre-natal second time	Delivery time	Post-natal first time	Post-natal second time	Non adopters	Total
<1	190	100	65	77	83	36	10	200
1–3	120	75	49	40	52	18	80	200
3–5	80	20	14	24	20	10	120	200
>5	10	7	5	4	8	3	190	200
Total	400	202	133	145	163	67	400	800

Source: Based on personal survey, 2010

Table 7.12 Caste/religion wise respondent's opinion regarding utilization of health care facilities

Religion/caste	Treatment	MCH	Family planning	Disease control	Vaccination	Laboratory	Total respondents
Hindu							
Upper caste	42	48	39	174	128	2	176
Backward	180	205	198	41	265	3	324
SC/ST	113	119	117	54	149	8	215
Muslim	45	28	10	24	50	1	85
Total	380	400	364	136	592	14	800

Source: Based on personal survey, 2010

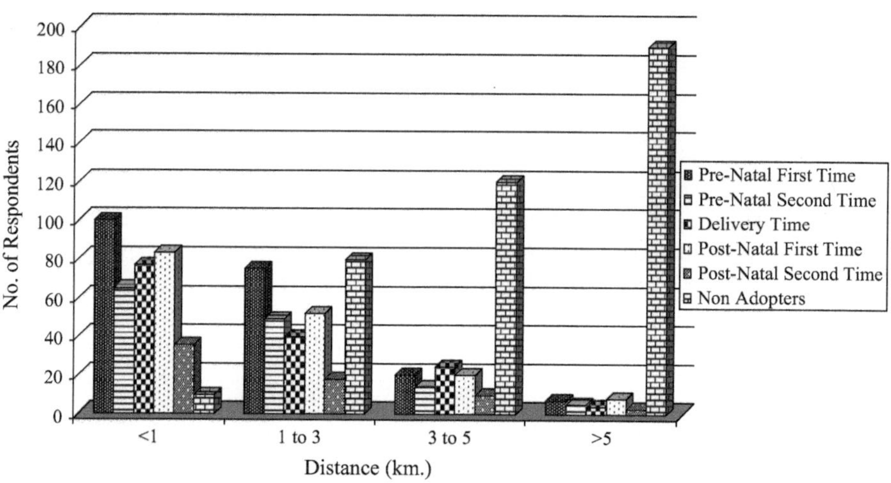

Fig. 7.4 Distance-wise respondents opinion in rural area of Varanasi district regarding utilization of MCH facilities, 2010

Post-natal second time help were received by less number of respondents in all the distance ranges. It is very well identified that distance factor has affected the number of medical advice seekers. In this study, it is clear that out of 400 registered cases for MCH only 16.75 % cases from renowned distance ranges had paid their visits for post-natal second time help. The frequency of non-users increase with increasing distance from PHC's. It is ostensible from Table 7.11 that number of medical advice seekers in second stage of pre and post-natal cases are less as distance increase. It displays that they are very much concern about the mother and child health care (Rai and Nathawat 2014).

7.4.1.4 Distance and Vaccination

The vaccination facility is most widespread among the PHC's facilities. As such maximum respondents always use this service. Though, the role of distance cannot be totally ruled out on the use of vaccination facilities. Out of 800 respondents, 592 persons (74 %) have used this facility. Among them maximum (198 persons: 33 %) are from less than 1 km distance followed by 1–3 km (180), 3–5 km (124) and above 5 km (90) distance ranges (Rai and Nathawat 2014).

7.4.1.5 Distance and Disease Control

Among the surveyed respondents only 136 respondents (17 %) from have visited PHC's in relation to disease control facility. In which maximum cases (70) had come from less than 1 km distance, 39 from 1 to 3 km, 21 from 3 to 5 km and only 6 cases had come after travelling more than 5 km distance (Rai and Nathawat 2014).

7.4.1.6 Distance and Lab Facilities

In general, PHC's have very poor lab facilities. Out of 800 respondents, only 14 cases were recorded for using lab facilities. Only block level PHC and CHC provide test facility and that too for malaria parasite test in blood and AFB (Acid Fast Bacilli) in sputum. There is no facility for X-ray and other pathological investigation. Thus, from the foregoing analysis it can be concluded that the hypothesis related to distance factor is confirmed (Rai and Nathawat 2014).

7.4.2 Caste/Religion-Wise Respondent's Opinion Regarding Health Care Facilities

7.4.2.1 Caste/Religion-Wise Respondent's Opinion Regarding Utilization of Health Care Facilities in PHC's/CHC's

The caste and religion are also considered important factors in affecting the utilization of health care facilities. In Indian set-up, caste not only reflects the social status but it also reveals economic status which is turn affect the user's interest

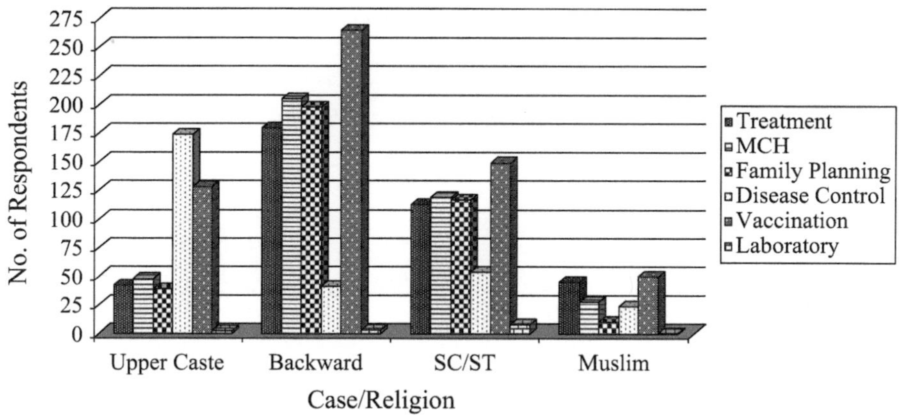

Fig. 7.5 Caste/religion-wise respondents opinion in rural area of Varanasi district regarding utilization of health care facilities, 2010

compulsions in availing the facilities. The area is inhabited by mainly two religious groups i.e. Hindu and Muslim. For the sake of convenience here, Hindus have been broadly classified into three major castes such as upper caste, backward caste and SC/ST. The caste/religion wise utilization pattern is presented in Table 7.12 which shows that the Hindus have adopted family planning service in higher proportion than their Muslim counterpart (Rai and Nathawat 2014). The utilization of MCH facility in Muslim is also found in lower proportion (Fig. 7.5).

Among the health care services, the utilization of vaccination ranks at number one in all the castes/religion. In upper caste and backward caste respondents MCH occupies second place after vaccination. But in Schedule Caste/Schedule Tribes (SC/ST's) and Muslim treatment is second choice. Comprehensive pattern of caste/religion wise utilization of different facilities reduced by PHC's is given in result (Rai and Nathawat 2014).

7.4.2.2 Caste/Religion-Wise Treatment Facilities in PHC's/CHC's

Treatment facilities are delivered only by PHC's/CHC's due to the sequence of experienced doctors. Doctors are found only on PHC's and district government hospitals. From the view point of caste the maximum utilization of treatment facility is found in SC/ST (52.55 %) and backward caste (55.55 %). It shows that the poor people visit PHC's/CHC's more regularly for utilization of services available there in. Caste/religion wise respondent's opinion regarding utilization of treatment facilities at PHC's /CHC's is shown in Table 7.13 and Fig. 7.6. Poor people cannot pay for the fees of private doctors and medicines advised by them and so they depend more on PHC's/CHC's (Rai and Nathawat 2014).

Table 7.13 Caste/religion wise respondents opinion regarding utilization of treatments facilities

Religion/caste	Total treated	Fever	Diarrhoea	Injury	Others	Untreated	Total respondents
Hindu							
Upper caste	42	14	12	7	9	134	176
Backward	180	48	57	24	48	144	324
SC/ST	113	25	42	21	25	102	215
Muslim	45	18	9	10	8	40	85
Total	380	105	120	65	90	420	800

Source: Based on personal survey, 2010

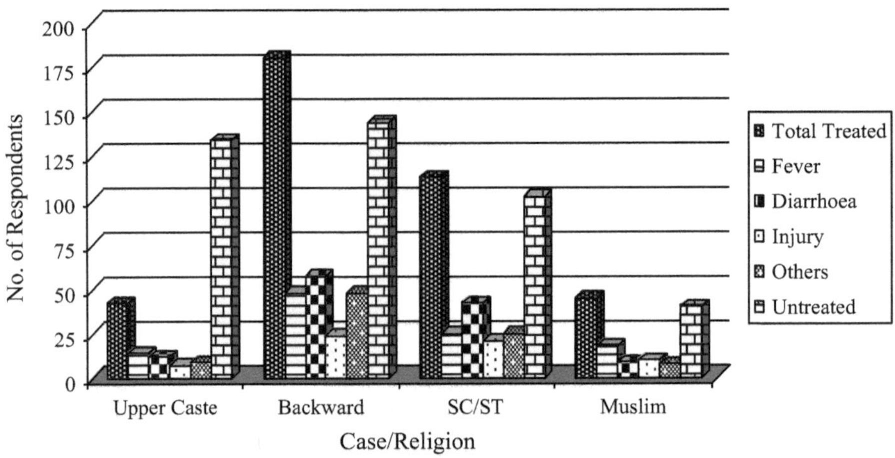

Fig. 7.6 Caste/religion-wise respondents opinion in rural area of Varanasi district regarding utilization of treatment facilities, 2010

7.4.2.3 Caste/Religion-Wise Utilization of Mother Child Health (MCH) Facilities

The people belonging to different castes and religions obtain MCH facilities in different ratio delivered by PHC/CHC and sub-centres (Table 7.14). Figure 7.7 shows that backward caste (64.81 %) people have utilized MCH service more than other castes. It is clearly seen in very low proportion in Muslim society (32.94 %). It designates that in Muslim community the considerable attention is not given on the mother's health (Rai and Nathawat 2014).

Looking at the caste/religion-wise data of pre-natal (before birth) and post-natal (after birth) first and second visit, it is easily seen that there is no consistency found in utilization of MCH services. It is notable to mention that upper caste people given more inclinations to PHC's/CHC's for delivery as compared to other castes. But in overall use of MCH facilities including pre-natal and post-natal consultation, it is

Table 7.14 Caste/religion-wise respondents opinion regarding utilization of MCH facilities

Religion/ caste	Total no. of MCH cases	Pre-natal first time	Pre-natal second time	Delivery time	Post -natal first time	Post -natal second time	Non adopters	Total respondents
Hindu								
Upper caste	48	10	35	25	18	9	128	176
Backward	210	112	60	81	96	32	114	324
SC/ST	114	64	27	29	35	21	101	215
Muslim	28	16	11	10	14	5	57	85
Total	400	202	133	145	163	67	400	800

Source: Based on personal survey, 2010

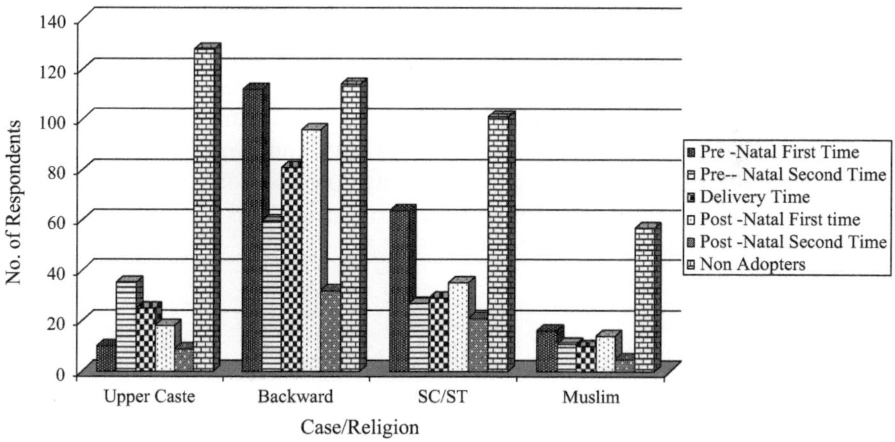

Fig. 7.7 Caste/religion-wise respondents opinion in rural area of the Varanasi district regarding utilization of MCH facilities, 2010

seen lowest in Muslim (32.94 %) than Hindu (52.02 %). Among Hindus it is perceived lowest in upper caste followed by SC/ST. The reason of lower use of MCH facility by upper caste people is not their bad intention towards mother and child health care; as an alternative they use better service obtainable in private hospitals in urban area (Rai and Nathawat 2014).

7.4.2.4 Caste/Religion and Vaccination Facility

Generally utilization of vaccination facility is found higher in all the castes/religious groups. Amongst the Hindus retrograde caste people have shown (81.79 % of total respondents) more attention in taking the advantage of comparison to other castes.

It displays that vaccination programmes are more famous among the peoples. It also specifies that people are well aware about profits and consequences of the various vaccination programmes (Rai and Nathawat 2014).

7.4.2.5 Caste/Religion and Disease Control

The utilization of disease control facilities is found maximum among Muslim and SC/ST. The use of this facility is found in lesser proportion in upper and backward caste. The social-economic condition of these castes is better than the Muslim and SC/ST people. Besides, they are more aware towards health (Rai and Nathawat 2014).

7.4.2.6 Caste/Religion and Lab Facilities

The utilization of lab facilities is poor in all castes and religions. Out of 800 surveyed people, only 14 persons have utilized lab facilities. Among them 8 persons belong to SC/ST, three persons from back ward caste and two persons from upper caste. The reason of poor utilization of lab facility does not indicate their indifferent attitude to it. Instead, it is due to almost non-existence of this facility out majority of PHC's/CHC's (Rai and Nathawat 2014).

7.4.3 Education-Wise Respondents Opinion Regarding Health Care Facilities

7.4.3.1 Education-Wise Utilization of Health Care Facilities

Education teaches the consciousness and awareness chooses the level of utilization of a particular facility. With this view it is tried to see the consequence of education (level of literacy) on adoption pattern. Out of 800 respondents whose opinions have been required, 22.5 % (180) are illiterate. It is marked from the Table 7.15 that majority of the respondents (97.4 %) have utilized the facilities of PHC's/CHC's and sub-centres for vaccination. Among the literates adoption rate was 77.41 %. In context to MCH, treatment and family planning, the percentage of illiterate's peoples have been 40.55 %, 47.77 % and 37.77 % respectively (Fig. 7.8). The percentage of well-educated respondents using MCH, treatment and family planning programmes comes about 81.79 %, 77.36 % and 81.31 % respectively. This obviously favors the hypothesis that more the education higher the utilization of health care facilities. Additional, the result of the survey shows another interesting detail that higher the level of education more the use of vaccination having primary,

Table 7.15 Education-wise respondents opinion regarding utilization of health care facilities

Education	Treatment	MCH	Family planning	Disease control	Vaccination	Laboratory	Total respondents
Illiterate	86	73	68	44	112	8	180
Literate							
Primary	106	82	64	62	152	4	206
Middle/ intermediate	164	174	180	18	242	2	310
Higher education	24	71	52	12	86	0	104
Total	380	400	364	136	592	14	800

Source: Based on personal survey, 2010

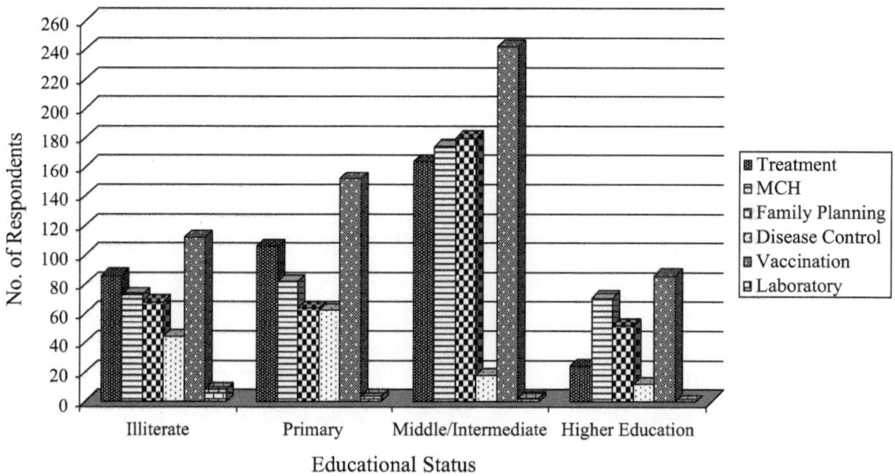

Fig. 7.8 Education-wise respondents opinion in rural area of Varanasi district regarding utilization of health care facilities, 2010

middle and higher education level account for 73.78 %, 78.06 % and 82.06 % respectively. It may be well-known here that the high percentage of utilization of health care facilities by the respondents literacy up to intermediate level is due to only their higher share in the sample (Rai and Nathawat 2014).

7.4.3.2 Education-Wise Utilization of Treatment Facilities

Table 7.16 shows the education-wise utilization of treatment facility. Among uneducated maximum use of PHC's/CHC's and sub-centres services has been made for treatment followed by fever (Fig. 7.9). The related trend is also observed in the case of educated respondents. Table 7.17 shows the education wise utilization pattern of MCH facilities. This service is used for checkup before (pre-natal) and after birth (post-natal) as well as counsel and delivery. Generally the respondents of the

Table 7.16 Education-wise respondent's opinion regarding utilization of treatment facilities

Education	Total treated	Fever	Diarrhoea	Injury	Other	Untreated	Total respondents
Illiterate	86	22	29	15	20	94	180
Literate							
Primary	106	29	31	17	29	100	206
Middle/ Intermediate	164	49	55	28	35	146	310
Higher education	24	8	5	6	6	80	104
Total	380	105	120	65	90	420	800

Source: Based on personal survey, 2010

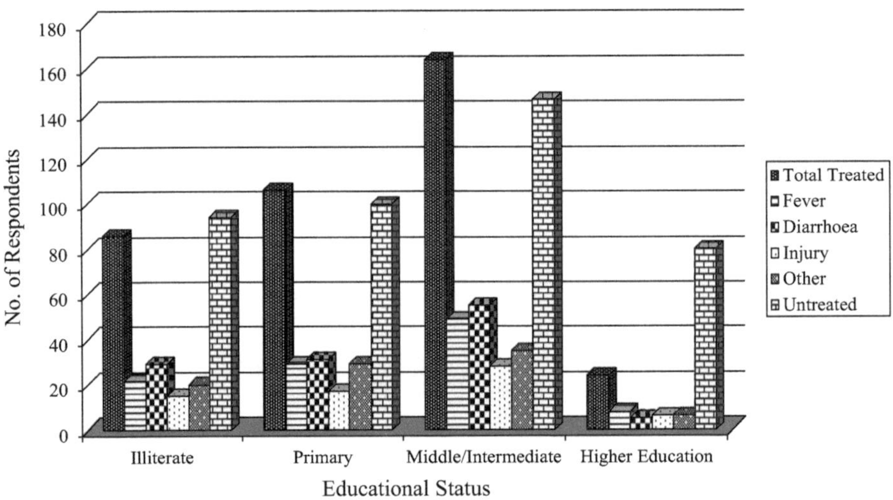

Fig. 7.9 Education-wise respondents opinion in rural area of Varanasi district regarding utilization of treatment facilities, 2010

Table 7.17 Education-wise respondent's opinion regarding utilization of MCH facilities

Education	Total no. of MCH cases	Pre natal first time	Pre natal second time	Delivery time	Post natal first time	Post natal second time	Non adopters	Total respondents
Illiterate	73	33	4	13	9	5	107	180
Literate								
Primary	82	58	22	22	26	10	124	206
Middle/intermediate	174	94	71	58	80	31	136	310
Higher education	71	17	36	52	48	21	33	104
Total	400	202	133	145	163	67	400	800

Source: Based on personal survey, 2010

Varanasi have utilized MCH facility accessible at PHC's and sub-centres double before and after natal (Rai and Nathawat 2014).

7.4.3.3 Education-Wise Utilization of Mother Child Health (MCH) Facilities

Table 7.17 obviously discloses that first pre and post-natal check-up and advice are realized more common than second time pre and post-natal advice. This trend is not healthy for the health of both mother and child. The facilities of PHC's/CHC's and sub-centres have also been used for delivery but in low quantity (Fig. 7.10). The percentage users is only 18.12 %, which is much more below from the expectation because PHC's and sub-centres have been destined to achieve the aim of safe delivery in rural areas. Therefore, it needs due to consideration and propagation (Rai and Nathawat 2014).

7.4.4 *House Type-Wise Respondents Opinion Regarding Health Care Facilities*

7.4.4.1 House Type-Wise Respondents Opinion Regarding Utilization of Health Care Facilities

In order to see the economic status wise utilization of health care facility house type has been selected to signify the economic situation of the country side people. Usually, the building materials used by the people indicate the economic well-being of the people. Out of 800 respondents, 592 persons have used vaccination facility

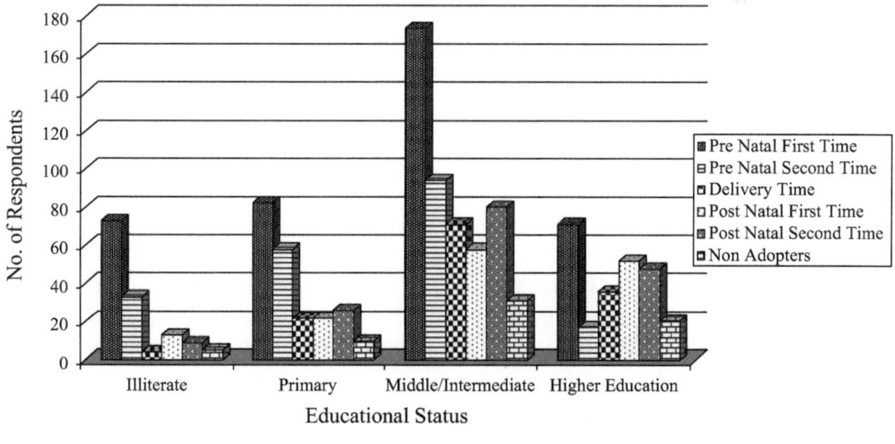

Fig. 7.10 Education-wise respondents opinion in rural area of Varanasi district regarding utilization of MCH facilities, 2010

Table 7.18 House type wise respondents opinion regarding utilization of HCF facilities

Types of houses	Treatment	MCH	Family planning	Disease control	Vaccination	Laboratory	Total respondents
Kachha (Non-Cemented)	150	100	364	72	162	10	266
Pucca (Cemented)	104	142	138	30	204	3	275
Mixed	126	158	124	34	226	1	259
Total	380	400	102	136	592	14	800

Source: Based on personal survey, 2010

Table 7.19 House type wise respondents opinion regarding utilization of treatment facilities

Types of houses	Treatment	MCH	Family planning	Disease control	Vaccination	Laboratory	Total respondents
Kachha (Non-Cemented)	150	100	364	72	162	10	266
Pucca (Cemented)	104	142	138	30	204	3	275
Mixed	126	158	124	34	226	1	259
Total	380	400	102	136	592	14	800

Source: Based on personal survey, 2010

Table 7.20 House types wise regarding utilization of MCH facilities

Types of houses	Total no. of MCH cases	Pre natal first time	Pre natal second time	Delivery time	Post natal first time	Post natal second time	Non adopters	Total respondents
Kachha (Non-Cemented)	100	65	36	46	43	14	166	266
Pucca (Cemented)	142	54	43	32	48	29	133	275
Mixed	158	83	54	67	82	24	101	259
Total	400	202	133	142	163	67	400	800

Source: Based on personal survey, 2010

obtainable at PHC's/CHC's. Among these users, pucca (cemented) households found 34 % while kachha (non-cemented) and mixed type households share 27.36 and 38.17 % respectively (Table 7.18). This means relatively well of respondents (72.62 %), utilize vaccination facilities more than frugally poor one.

It is remarkable to note that pucca (cemented) and mixed (kachha-pucca) house landlords appear to be more cautious for the health of mother and child and also about the family size. They found about 75 % and 71.97 % respectively (Rai and

Nathawat 2014). This confirms that there is definite constructive relationship between the economic condition and use of services. Better the economic situation, more the use of health care services and vice-versa. With regard to utilization of treatment, illness control, and lab services from PHC's/CHC's the share of poor people having kachha (non-cemented) houses is more i.e. 39.47 %, 52.94 % and 71.42 % respectively in contrast to frugally well of people having pucca (cemented) as well as mixed houses. This designates that poor people are more dependent on facilities rendered by PHC's/CHC's. Their frugalities do not permit to avail costly medical advice and treatment in city area (Rai and Nathawat 2014).

Table 7.19 exposes the breakup of respondents according to their visits for using treatment facilities. The rural people have used PHC's/CHC's services mostly for the fever, diarrhea and other injury. The influence of diarrhea, fever, injury and other services users is 31.57 %, 27.63 %, 17.10 % and 23.68 % respectively. It is marked from Table 7.19 that the kachha (non-cemented) house proprietors shared fairly large proportion i.e. 44.76 %, 48.33 % and 32.30 % amongst the total fever, diarrhea and injury service respondents (Rai and Nathawat 2014).

Table 7.20 clearly shows house wise utilization trend of MCH facilities available at PHC's/CHC's and sub-centres. It is ostensible from the study that the utilization of MCH facilities by poor people having kachha house users share about 25 % amongst the total users of this service (Rai and Nathawat 2014). In MCH facilities the users have required the assistance of doctors/other para medical staffs for pre-natal (before birth), post-natal (after birth) and delivery advices. It may be well-known here that good numbers of people have not wanted the pre and post-natal services second time. It is a worrying trend so far the health of mother and child concerned. They have used the pre-natal and post-natal facilities double before and after birth of their children. The proportion of respondents having kachha (non-cemented) house has been seen as 30.10 %, 24.78 % and 31.78 % with respect to pre and post-natal delivery facilities. This clearly designates that poor respondents pay less significant care on the health of mother and child (Rai and Nathawat 2014).

7.4.5 Occupation-Wise Respondents Opinion Regarding Utilization of Health Care Services

Occupation of a person delivers an impression about the income condition as well educational status. As such occupation wise respondent's opinion regarding utilization of health care services has been required. The results have been given in Table 7.21 and Fig. 7.11. It is very significant to note that those who are either in jobs or supervise agriculture along with services they use less health care facilities obtainable at PHC's and sub-centres. The cause is very simple. They are frugally well so they can pay for private doctor's services as well as specialized medical services available in urban area. Besides, they are aware sufficient in utilizing better health facilities for their illness (Rai and Nathawat 2014).

Table 7.21 Occupation type wise respondents opinion regarding utilization of health care facilities

Occupation	Treatment	MCH	Family planning	Disease control	Vaccination	Laboratory	Total respondents
Labourers	146	124	114	66	224	8	250
Cultivators	130	128	126	28	238	3	266
Agriculture based Cottage Industry	59	97	61	16	62	1	130
Agriculture with service	31	20	69	14	40	2	89
Others	14	31	24	12	28	–	65
Total	380	400	364	136	592	14	800

Source: Based on personal survey, 2009–2010

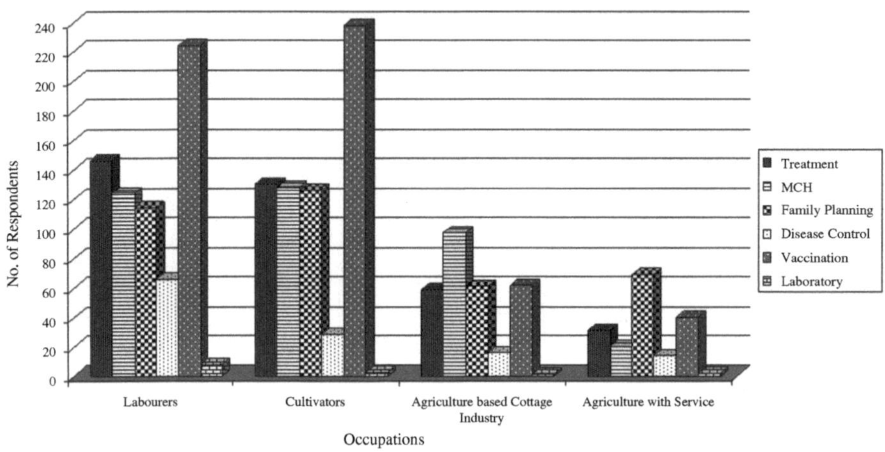

Fig. 7.11 Occupation-wise respondents opinion regarding utilization of health care services in Varanasi, 2010

Their percentage for treatment MCH, family planning, disease control and vaccination are about 11.84 %, 10.25 %, 17.30 %, 19.17 % and 11.48 % respectively. Against this, laborer's and poor farmers depend more on health care facilities obtainable at PHC's. For the prior study it is perceived that due to numerous reasons the pace of utilization of health care facilities in the study area is inactive and in certain cases it is quite low. As well, poor people depend more on PHC's and their sub-centres. Therefore, more care should be made to equip them well efficiently (Rai and Nathawat 2014).

Table 7.22 Utilization pattern of the health care facilities

Health care facility	No. of respondents availing the facility	%
Private practitioners	327	40.88
Government hospital	313	39.12
Traditional practitioners	83	10.37
Home remedy	45	5.63
Others	32	4.0
Total	800	100.00

Source: Based on personal survey, 2009–2010

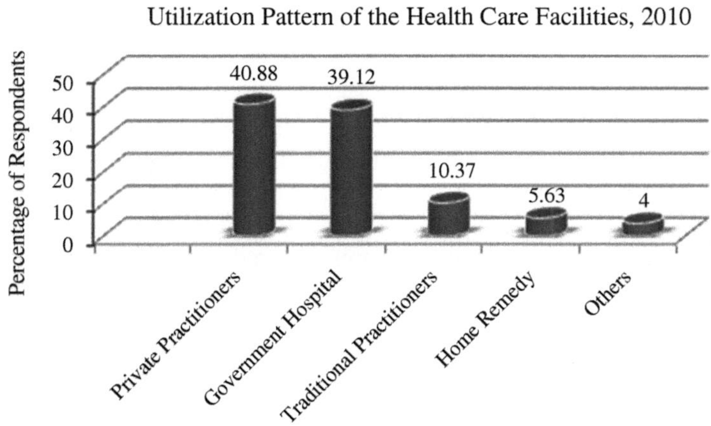

Fig. 7.12 Utilization pattern of the health care facilities

7.4.6 Nature of Health Care Facilities and Their Availability

There are different kinds of health facilities available in the study-area. Utilization pattern differs on a number of aspects like travel distance to use the services, cost of treatment, and popularity of resource persons employed there etc. To know the order of preference towards using a specific health care service, respondents were demanded to give their favored order and outcome so achieved is presented in Table 7.22 and Fig. 7.12.

Table 7.22 highlights that maximum numbers of respondents (40.88 %) are utilizing the facilities rendered by private medical practitioners. It is perhaps due to facilities available there and proper care and interest taken by private doctors. About 39.12 of the total respondents mostly belonging to low income group visit to government hospital. Out of the total respondents approximately 10.37 % avail the traditional practitioners while 5.63 % people believe in home-remedy. These respondents belong to poor background of the society having low literacy rate and have more faith in *Ojhas* and *Vaidyas*. It has been distinguished during the survey that many of the respondents prefer to go hospitals for specific purpose such as immunization, MCH, family planning etc. (Rai and Nathawat 2014)

7.4.6.1 Causes of Giving Preference to PHC's/New PHC's/CHC's/Sub Centres

With vast asset and availability of trained and experts doctors at PHC's and Government hospitals these were ranked at second order of preference. Therefore, it is important to evaluate people's attitude towards functioning of government hospitals or PHC's/New PHC's/CHC's/Sub Centres. Table 7.23 shows that maximum number of respondents (42.5 %) makes visit to PHC's/CHC's for immunization of their children or with pregnant women, because these facilities are delivered by the government hospitals without any costs (Rai and Nathawat 2014). About 25.25 % of respondents visit governments hospitals on account of very low costs of treatment while 11.63 % people come to these centres due to distance factor. Further, a very small proportion of respondents i.e. 5.87 % visit to PHC's /CHC's due to obtainability of free medicines (Fig. 7.13).

From the above investigation it is very clear that health care services delivered by the Government like PHC's or Hospitals are not still utilized appropriately because of the several, reason stated by respondents like irregularity in the availability of doctors, poor maintenance and non-availability of recommended medicines (Fig. 7.14).

Table 7.23 Reason of giving preference to PHC's/CHC's or government hospitals

Reason	No. of respondents	%
Easiest source of immunization	340	42.5
Near	93	11.63
Cheap	202	25.25
Near/Cheap	118	14.75
Free medicine	47	5.87
Total	800	100

Source: Based on personal survey, 2009–2010

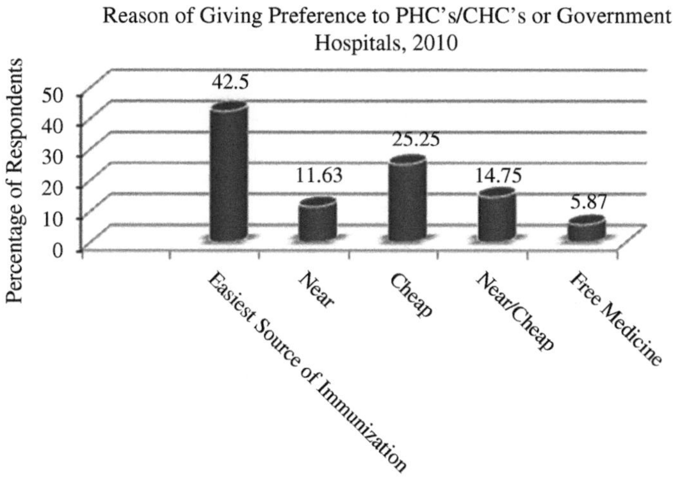

Fig. 7.13 Reason of giving preference to government health centres or hospitals

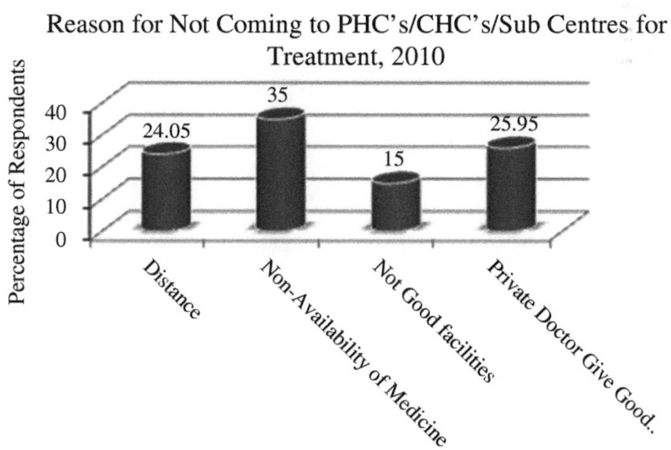

Fig. 7.14 Reason for not coming to government health centres

7.4.6.2 Causes for Not Coming to PHC's/New PHC's/CHC's

To find out the respondents approach towards not coming to PHC's/New PHC's/CHC's and other Government Hospitals responses were presented. Table 7.24 shows that maximum number of respondents (35.00 %) do not wish to visit PHC's/New PHC's/CHC's or Government Hospitals for treatment because of non-availability of appropriate remedies (Rai and Nathawat 2014). About 24.05 % of respondents do not avail the facilities delivered by government at PHC's/New PHC's/CHC's and hospitals because of distance factor (Rai and Nathawat 2014).

Table 7.24 Reason for not coming to PHC's/CHC's/sub centres for treatment

Reason	No. of respondents	%
Distance	101	24.05
Non-Availability of medicine	147	35.00
Not good facilities	63	15.00
Private doctor give good service	109	25.95
Total	420	100.00

Source: Based on personal survey, 2009–2010

Table 7.25 Level of satisfaction of respondents

Level of satisfaction	No. of respondents	%
Satisfied	203	25.38
Partially satisfied	480	60.00
Not satisfied	117	14.62
Total	800	100.00

Source: Based on personal survey, 2010

Fig. 7.15 Level of satisfaction of respondents

Out of the total respondents nearly 25.95 % choose to go their health services to private medical consultants because of their good services and availability for all the time. Pathological and important services are not well maintained at the Government hospitals therefore approximately 15 % of the respondents do not visit these centres (Fig. 7.14).

They favor to go at PHC's/New PHC's/CHC's only for specific facilities like immunization, MCH etc. (Fig. 7.12).

7.4.6.3 Level of Satisfaction of the Respondents

To find out the level of satisfaction from efficacy and facilities provided by government PHC's/New PHC's/CHC's and hospitals respondents were again wished to give their response mainly on this issue. Their answers were considered into three level of satisfaction i.e. satisfied, partially satisfied and not satisfied. Frequency of respondents with these levels of gratification is obtainable in Table 7.25 and Fig. 7.15. This highlights that a large number of respondents (60 %) are partly pleased with efficacy and medical facilities available at PHC's and other Government hospitals (Rai and Nathawat 2014).

It means that services rendered by PHC's/New PHC's/CHC's are neither poor nor so good. About 25.38 % respondents are fully gratified with the services available at these PHC's. Out of the total 800 respondents 117(14.62 %) are not satisfied with the operative of PHC's. Most of them have grievance towards the non-obtainability of doctors and medicines both at the PHC's and Government hospitals because of their assignation in private practices somewhere else (Table 7.25). Thus here is necessity to improve better services of health care at PHC's/New PHC's/ CHC' and sub-centres of rural area of Varanasi district (Rai and Nathawat 2014).

References

Aday, L. A., & Andersen, R. (1974). A framework for the study of access to medical care. *Health Service Research, 9*, 208–220.

Akhtar, R., & Khan, A. Q. (1993). Spatial organization of health facilities in Jammu and Kashmir. *Annals NAGI, 13*(2), 29–38.

Basu, R. N. (1984). Immunization programme in comprehensive health care. *Indian Journal of Medical Education, 13*, 47–49.

Datta, S. P., & Kale, R. V. (1969). An operation research study in primary medical care in Pondichery. *Indian Journal of Preventive Social Medicine, 1*, 65–72.

Ghosh, B. N., & Mukharjee, A. B. (1989). An analysis of health services coverage of primary health centre in West Bengal. *Indian Journal of Public Health, 33*(1), 12–19.

Khan, M. E. (1980). *People's perception about family planning in India: A study of Andhra Pradesh and Bihar*. New Delhi: Concept Publishing Co.

Kumra, V. K., & Singh, J. (1994). Assessment of Community Attitude regarding the services of primary health centre: A medical geography study. *Indian Journal of Preventive & Social Medicine, 25*(3/4), 118–126.

McPhail, T. E. B., Wilson, E. E. L., & Eckersley, L. W. (1963). A study on the working of primary health centers in Uttar Pradesh (India), WHO/SEA/RH/14.

Mishra, C. P. (1988). A study of community practices regarding MCH care in rural areas of Varanasi. *India Journal of Public Health, 32*(2), 92.

Mohler, H. (1978). *Primary health care justice in health* (p. 3). Geneva: World Health.

Murali, I. (1981). Pattern of utilization of health care facilities by a selected urban community in New Delhi. *Indian Journal of Preventive and Social Medicine, 12*(2), 117–125.

Prakasam, C. P. (1995). Utilization of primary health care services in a rural setup. *Health Population Perspective Issues, 18*(2), 72–80.

Rai, P. K., & Nathawat, M. S., (2014). Utilization of health care services in Varanasi District: A geographical analysis, GEOGRAFIA, Malaysian Journal of Society and Space (online), 10(2), 14–33. (ISSN: 2180–2491)

Sinha, S., & Rajeswari. (1993). Social structure and utilization of health care facilities in rural Haryana. *Annal NAGI, 30*, 37–47.

Srinivasan, S. (1984). A primary health centre in Tamil Nadu, Kurukshetra, p. 15.

Chapter 8
GIS Initiatives in Health Care Planning

Abstract The geographical information system (GIS) is very helpful in a variety of application areas points to an increasing interest in the spatial aspects of health policies and planning. The main aim of this chapter is to examine the relevance of Geographical information system (GIS) supporting health planners for a district level healthcare planning. For this purpose, an effort has been made here to calculate the hospital requirement area to know the specific sector that needs to better develop health facilities. The weightage is assigned to the class of thematic layers respectively to produce weighted thematic maps, which have been overlaid and numerically added in order to produce a Hospital requirement index (HRI) and hospital requirement zone (HRZ) map. These maps are very useful to calculate the exact area having good health facilities and also those wherein healthcare facilities need to be improved in Varanasi district. When the outcomes are compared with Multiple Linear Regressions and Information Value method then it is found that the malaria model developed with Information Value method is an optimum model which is selected for the measurement of hospital requirement index using the related parameters. The Hospital requirement index (HRI) values according to the weighting method are found to lie in the range from 11 to 23. After calculation by weighting method using selected indicators, it is found that the areas coming under very high and high requirement class is 46.62 % and 7.55 %, respectively, whereas 3.39 % and 42.63 % of the total areas comes under low and moderate requirement classes in Varanasi district.

Keywords GIS • Remote sensing • Hospital Requirement Index (HRI) and Hospital Requirement Zone (HRZ) • Multiple linear regressions and information value method

8.1 Introduction

GIS and associated spatial analysis approaches deliver a set of tools for defining and understanding the changing spatial group of healthcare for exploratory its relationship to health outcomes and access, and for exploring how the delivery of healthcare can be enhanced (Rohan 2002; Rai et al. 2011; Rai and Nathawat 2013). It

P.K. Rai, M.S. Nathawat, *Geoinformatics in Health Facility Analysis*,
DOI 10.1007/978-3-319-44624-0_8

deliberates the use of GIS in analyzing healthcare necessity, access and utilization, planning and assessing service locations and also in spatial decision support for healthcare distribution. The acceptance of GIS by healthcare researchers and policy-makers will depend on access to integrated spatial data on health services utilization and results connected to human service systems (Richards et al. 1999; Rai et al. 2011; Rai and Nathawat 2013). Healthcare covers wide-ranging spectrum extending from personal health facilities to health education and information for preclusion of diseases, timely diagnosis, treatment and rehabilitation (Gatrell and Loytonen 1998; Gatrell 2002; Rai et al. 2011; Rai and Nathawat 2013). The significance of GIS is appropriate to the aim of modern public health which has been set by World Health Organization (WHO) as "the achievement by all people of the highest conceivable level of health".

Health has been avowed as a functional right; so the each state has a accountability to deliver good healthcare services to the people (Srinivasan 1984; Rai et al. 2011; Rai and Nathawat 2013). There are so many difficulties in delivery of healthcare services on account of which people find themselves incapable to utilize the services upto its satisfactory level. The major hitches connected with healthcare services and their utilization in the study area are uneven delivery of healthcare services, illiteracy, poor socio-economic conditions along with lack of consciousness to benefit the healthcare services that stand in the way of enhancement of health in the area. Utilization pattern of healthcare services obviously specifies the consciousness and attitude of people towards their health (Prakasam 1995; Rai et al. 2011; Rai and Nathawat 2013). Education, economy, male-female ratio and social status are main influencing factors for utilization of healthcare services. A literate person is more cautious about his health than an uneducated. Females utilize these facilities less as compared to males (Sinha and Rajeswari 1993). Besides, efficacious utilization of health facilities depends on dependability, motivation and finally on the awareness of the people about the services and the need about a specific facility (Kumra and Singh 1994; Rai et al. 2011; Rai and Nathawat 2013).

In India, the healthcare services in rural area are being provided through the network of primary health centres (PHCs) and related sub-centres (Dutt 1962; Datta and Kale 1969; Rai et al. 2011; Rai and Nathawat 2013).

The study goals at calculating hospital requirement index (HRI) and hospital requirement zone (HRZ) though satellite data and GIS techniques and utilization of health care services in Varanasi district to describe state-of-the-art in healthcare services and express the co-ordinated plans and strategies in the health sector.

8.2 Materials and Methodology

Various GIS themes on specific parameters which are connected to the occurrence of malaria i.e. land use, normalized difference vegetation index (NDVI), distance to water ponds, distance to river/stream, distance to road, distance to hospital, rainfall, temperature, and projected population density of year 2009 have been created using

Ilwis Version 3.4 and Arc GIS Version 9.3 and ERDAS Imagine Version 9.1 software (Rai et al. 2012, 2011; Rai and Nathawat 2013).

Statistical software SPSS Version-16 is used to develop the various layer maps that help in the preparation of the malaria susceptibility map using three statistical methods. Topographical map of 1:50,000 scale is used to extract district and development block boundaries. The coordinates of present healthcare services units are calculated during the field survey through Global Position Systems (GPS). The GIS vector maps like road network, water bodies, PHC's/hospitals locations etc. are prepared from the IRS-1C LISS-III remote sensing data of 2008. Thus, land use land cover (LULC) map, NDVI, GIS layer of water bodies and other important parameters used in this study are prepared in ERDAS imagine 9.1 and ARC GIS 9.3 software.

In this study optimal model is used to compute hospital requirement zone (HRZ) by heuristics (weighting) method (Rai et al. 2012). When the outcomes are compared with Multiple Linear Regressions and Information Value Method then it is found that the malaria model developed with Information Value Method is an optimum model which is selected for the measurement of hospital requirement index using the related parameters (Rai and Nathawat 2013).

The weighted linear combination (WLC) method is a decision rule for developing composite maps using GIS. It is one of the utmost frequently used decision models in GIS platform. The technique, however, is often applied without full understanding of the assumptions underlying this method. Weightage overlay method is being applied to compute the healthcare services susceptible areas followed by indexing to layers and these are very significant input for the preparation of hospital susceptible index (HIS) (Rai and Nathawat 2013). This has been done using raster layers conveying relative weightage in accordance to its effect and expert view; the weighted layers are overlaid to find out its potential in healthcare facilities. Simply by adding the weightages, hospital susceptible maps have been developed where a high weight relates to high susceptible area. Classes with high, moderate and low are generated to highlight the area that fall under high, medium and low susceptible zone in respect of present healthcare services and these information are very significant for healthcare services, management and planning purposes. Complete methodology is shown in Fig. 8.1.

The numerous data layers i.e. distance to district road map and distance to highway map etc. has been prepared in weightage values (from 1 to 5). Each class within a layer has been given a weightage value; the highest class has 5 values, the medium class 3 values and the lowest class has only 1 value. The weightage is assigned to the classes of each thematic layers respectively to produce weighted thematic maps, which have been overlaid and numerically added according to equation 1 to produce a hospital requirement index (HRI) map and this is very useful to calculate the exact area in the Varanasi district where health facilities is good and where health care services essential to be enhanced (Rai and Nathawat 2013).

$$HRI = Dro + Dh + Dhc + Dch + MSI \tag{8.1}$$

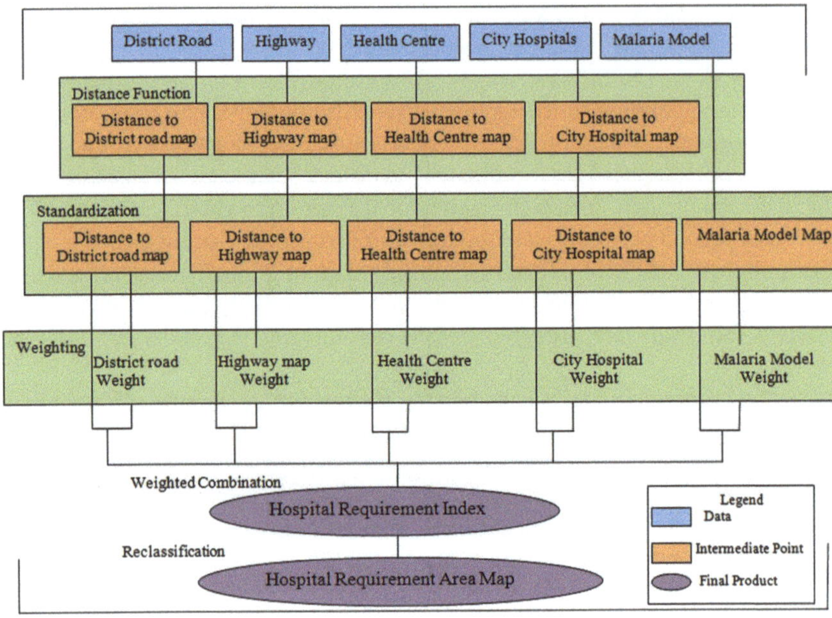

Fig. 8.1 Schematic representation of preparation of HRZ map in weighting model

where *Dro, Dh, Dhc, Dch, MSI* are distribution-derived weights for distance to road, distance to highway, distance to health centre, distance to city hospitals and malaria susceptibly index (MSI) model respectively.

Primary data is also collected from 800 respondents of 16 selected villages (2 villages from each development block) in the rural area of Varanasi to know about the utilization of healthcare services and their outcomes are studied with the help of SPSS software. City area of Varanasi is not measured for this purpose because here many good private and government hospitals are established where people find quite good health care services in comparison to the rural part of the Varanasi (Rai and Nathawat 2013).

8.3 Result & Discussion

8.3.1 Distance to Road & Highway and Weightage

Distance to district road has been developed to assigning the weightage (Rai and Nathawat 2013). Weightage is allocated on the basis of road distance from the health care services situated in the rural and urban area. Lower weightage (1) is given to the roads whose distance is <300 m from the health facilities where as higher weightage (5) given to roads having distance more than 3000 m from the health care

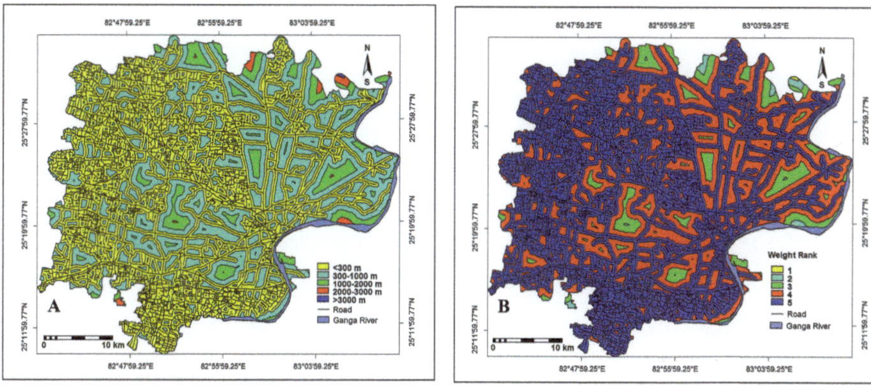

Fig. 8.2 Distance to road (**a**) and its weight rank (**b**)

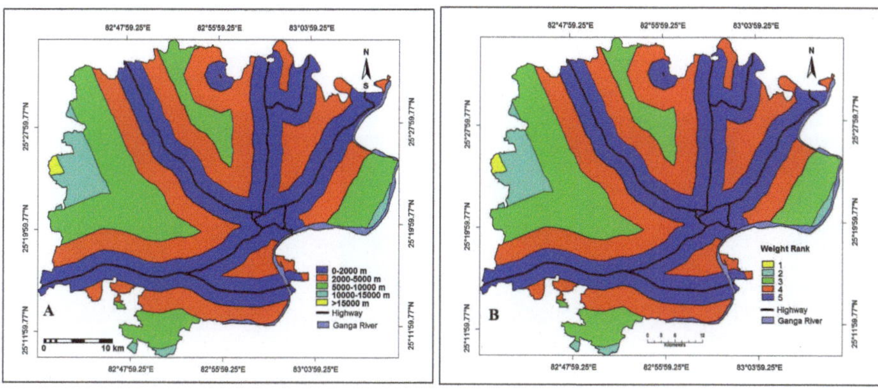

Fig. 8.3 Distance to highway (**a**) and its weight rank (**b**)

units (Figs. 8.2 and 8.3). In the same way, higher to lower weights are also given to the distance to highway map. Lower weightage (1) is given for class <2000 m, but for class >15,000 higher weightage (5) is allocated.

8.3.2 *Distance to Health Centres & City Hospitals and Weightage*

Distance to health centres and city hospitals are also calculated and their weightage are allocated on the basis of distance to the areas which is either distant or nearby from the primary health centres and city hospitals (Figs. 8.4 and 8.5). Weight rank 1 is given to the area, which comes very near (<1000 m) to these health centres and where health services are quit adequate while weightage 5 is assumed to the area being situated very far (>8000 m) from the present health centres (Rai and Nathawat

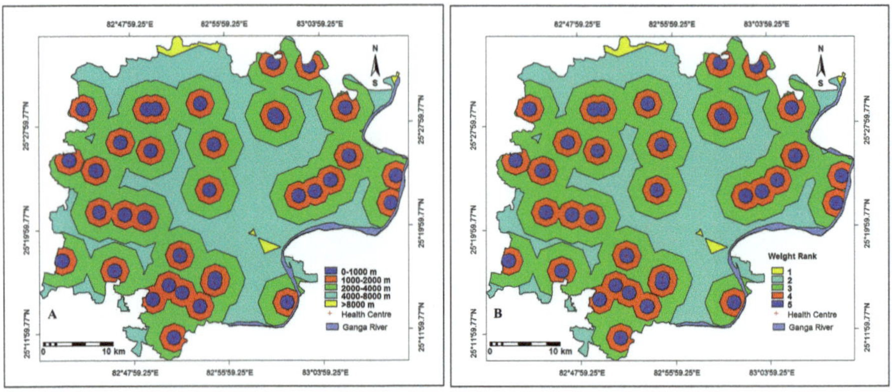

Fig. 8.4 Distance to PHCs/CHCs (**a**) and its weight rank (**b**)

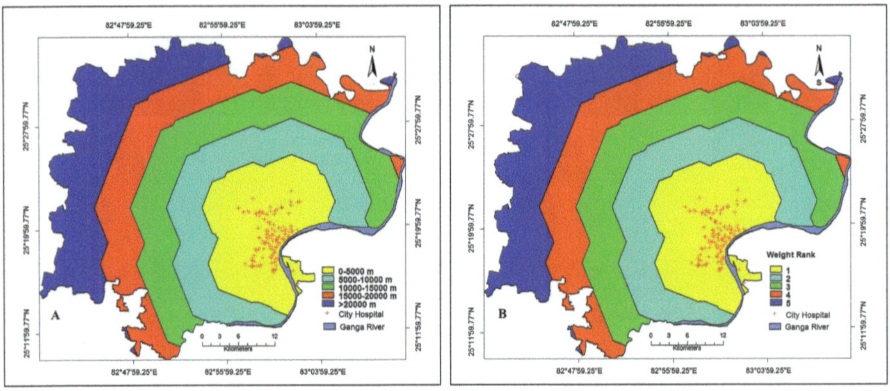

Fig. 8.5 Distance to city hospital (**a**) and its weight rank (**b**)

2013). People living in the area could not easily approach these health centres because distance becomes a main compel. People live in the north-eastern and south-east part of Varanasi district are very distant (>20,000 m) from the urban hospitals and they are generally reliant on the primary health centre, where health care facilities are not up to their wish. So that, this class is assigned higher weight rank (5), whereas weight rank 1 is allotted for those area, which comes in <5000 m buffer area from the city hospitals (Figs. 8.4 and 8.5).

8.3.3 Weightage of Malaria Susceptible Model Map

Malaria susceptible index is calculated and a malaria model map is prepared to estimate the malaria susceptible zones by information value method. In this, influential parameters measured in malaria mapping are: Rainfall (Rf), Temperature

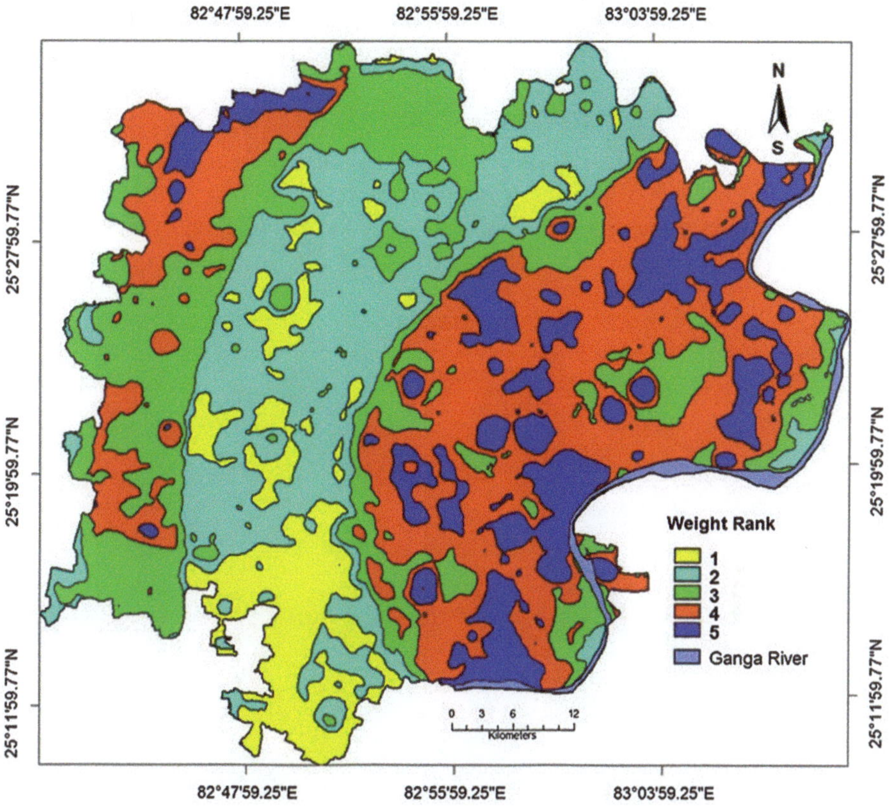

Fig. 8.6 Weight rank to malaria model (MSI model)

(Temp), Population density (Pd), Distance to river (Dri), Distance to road (Dro), Distance to health facilities (Dhf), Land use/Land cover (Lu/Lc) and Normalized Difference Vegetation Index (NDVI). These parameters were very obliging to identify its effect on increasing and decreasing area percentage of malaria disease. Distance to health services, population densities etc. became one of the significant factors for malaria disease mapping. Malaria models measured through information value itself became vital inputs to analyze the hospital requirement index (Rai and Nathawat 2013). Weight ranks are also allocated to malaria model, which is based on occurrence of malaria susceptibility zones in the area. Areas which fall under the higher susceptible malaria zone are given higher weightage 1 whereas low weightage 1 is allocated to very low susceptible malaria zone (Fig. 8.6).

8.4 Hospital Requirement Index (HRI) and Hospital Requirement Zone (HRZ)

The hospital requirement index (HSI) values from the weightage approach are seen to lie in the range from 11 to 23 (Fig. 8.7). The cumulative frequency curve of HRI values has been divided into four categories representing near equal delivery to yield four hospital requirement zones i.e. low, moderate, high and very high (Rai

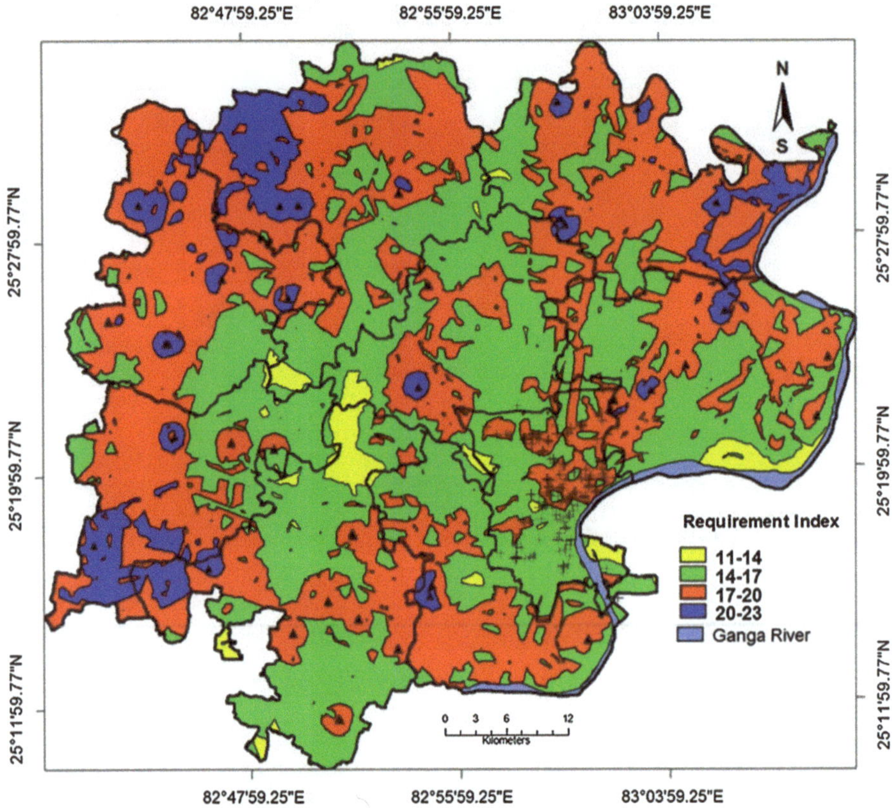

Fig. 8.7 Hospital Requirement Index (HRI) in Varanasi district

Table 8.1 Area statistics of hospital requirement class in Varanasi district

Requirement class	No. of pixel	Pixel percentage	Requirement area (%)	Area (sq.km.)
Low requirement	20,745	2.15	3.39	51.86
Moderate requirement	260,613	26.97	42.63	651.53
High requirement	283,758	29.36	46.42	709.39
Very high requirement	46,154	4.78	7.55	115.38

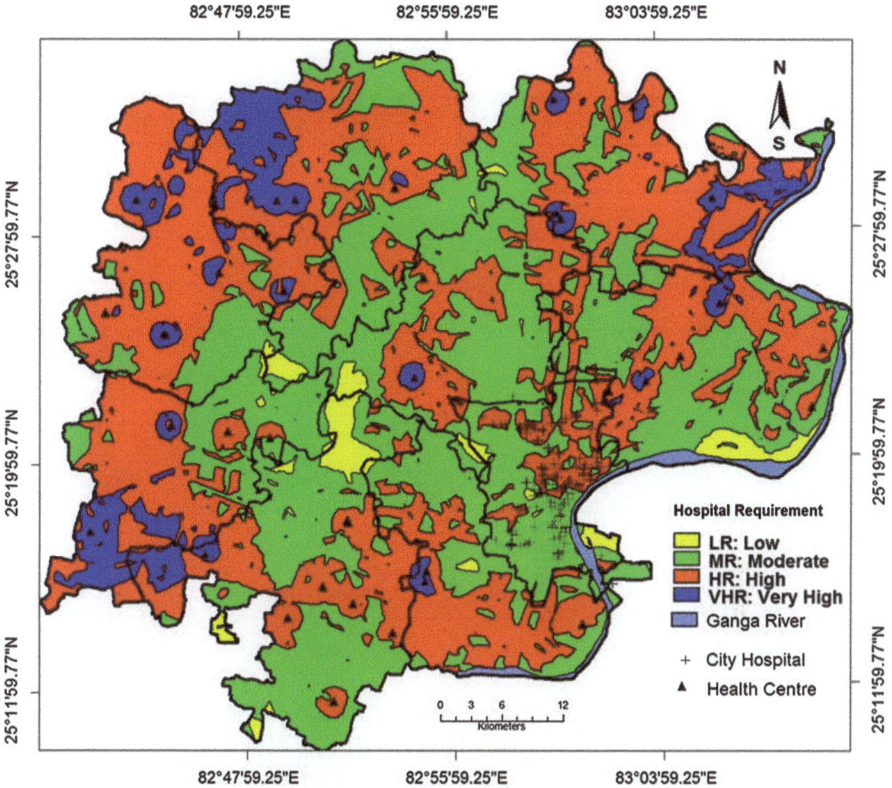

Fig. 8.8 Hospital Requirement Zone (HRZ) in Varanasi district

and Nathawat 2013). After calculation by weightage method using all the above selected parameters, it is clearly seen that the areas coming under very high and high requirement classes are 46.62 % and 7.55 % respectively whereas 3.39 % and 42.63 % of the total areas come under low and moderate requirement class in Varanasi district (Table 8.1, Figs. 8.8 and 8.9).

8.5 Reasons for Systematic Planning of Health Facilities

8.5.1 Inadequate Facilities and Network of Health Centres and Sub-Centres

The programme of founding primary health centres (PHCs) in each development block having a population between 60,000 and 80,000 was started as an essential part of the community development programme in 1952. Each primary health centre consisting of 6 beds were situated at the block headquarters and with these 4

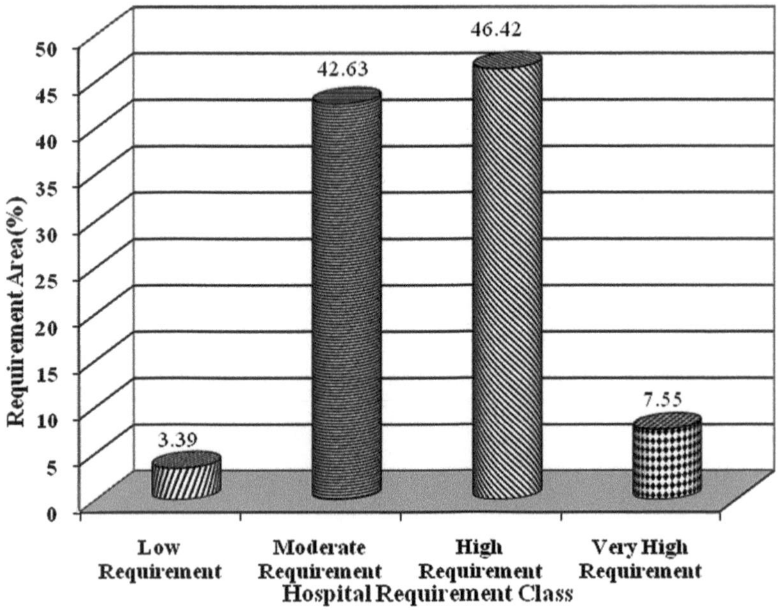

Fig. 8.9 Graphical representation of hospital requirement classes

sub-centres were attached. The staff comprises 1 medical officer, 1 sanitary inspector, 4 mid-wives (ANMs) and 2 additional personal. The centre was to be buoyed by district hospitals for referral consultation, laboratory, medical, surgical, nursing and administrative facilities (Rai and Nathawat 2013).

In the rural area, facilities are provided through a network of integrated health and family welfare distribution system. Health care programmes have been modernized and reoriented from time to time for achieving the objective 'Health for all' of the government and also of the National Health Policy (Rai and Nathawat 2013). Priority has been rendered to extension, expansion and consolidation of rural health infrastructure like sub-centre, primary health centres (PHCs) and community health centres (CHCs). Keeping the aforementioned objectives in view, it was obvious to have one primary health centre for every 30,000 population (20,000 population in hilly and tribal areas), a CHC on every 120,000 population (80,000 population in hilly and tribal areas). Until 1991, these facilities were extended through a network of 22,229 PHCs, 131,379 sub-centres and 1923 community health centres (CHCs). As such, the country got a primary health centre (PHC) for about 38,000 people and sub-centre for 6500 people, showing more burden than targeted standards. As for the growth of PHCs and sub-centres, states like Andhra Pradesh, Bihar, Gujrat, Karnataka, Madhya Pradesh, Maharastra, Tamilnadu, Uttar Pradesh, Haryana and Andaman and Nicobar Islands need more PHCs and sub-centres so as to achieve the request in healthcare sector in the future (Rai and Nathawat 2013).

Currently, in the area under study there are 3 CHCs, 32 PHCs and 234 sub-centres. In view of the standards given above, it is significant to evaluate the present condition of health care network (Rai and Nathawat 2013). The study area has a CHC for about 483,379 populations whereas the standard is much lower than this. In order to attain the demand of CHC, nine CHCs have been planned at Pindra, Gangapur, BariyaSanpur, Jalhupur, Harhua, Sewapuri, Kachnar, Shivdaspur and Sadalpura. Only one development block (Baragaon) completes the planned standard norm, while the other development blocks lag behind the standard norms (Rai and Nathawat 2013). In the study area, a PHC is found on a population of 45,316. Compared to the suggested norm, there appears to be a lack of 16 PHCs in the area. New PHCs have been premeditated at Rasulpur, Hiramanpur, Kashipur, Mangari, Pura Raghunath, Raund Khurd, Narainpur, Aigar, Lamhi, Mohaw, Kardhana, Bhikhampur, Benipur, Lohta, Dafi and Kotwar. If one looks for numerical situation of sub-centres in the area then there is one sub-centre on each 6197 persons (Rai and Nathawat 2013). Following this norm, an extra 56 sub-centre will be needed to provide improved health facilities in the area. It may be specified that while suggesting CHCs and PHCs, proper care of appropriate place has been taken in view (Rai and Nathawat 2013). Where it was thought suitable, existing PHCs and sub-centres have been progressed into CHCs and PHCs, respectively.

8.5.2 Poor Infrastructure Facilities at Primary Health Centres (PHCs)

Infrastructure states to the elementary support system in the form of property, maintained building and the basic services obtainable within the premise for smooth operative of the health care system. Such services include supply of water, electricity, laboratory service for testing blood, urine etc., telephone, functional vehicle, delivery room, injection and first-aid etc.

In the study area, the PHCs are treated as an important pillar for providing primary healthcare services to the rural masses in the Varanasi district and only 78.1 % PHCs of the study area have their own building. About 9.3 % and 12.5 % PHCs are running in Panchayat Bhawan and rented buildings respectively. For appropriate expansion of health services in Varanasi district, it is essential to deliver individual building to every PHCs and sub-centres (Rai and Nathawat 2013).

Regarding the supply of water, it is very apathetic discouraged to mention that only 40.6 % PHCs get consistent tap water supply. The remaining 59.4 % PHCs obtain water form hand pumps. At block level, considerable variation has been identified. Presence of overhead tank is essential for incessant water supply, but the majority of the PHCs are destitute of this service (Rai and Nathawat 2013). The continuous supply of electricity is an essential for smooth and effectual operative of health care services at PHCs. In the Varanasi district, about 87.5 % PHCs have only electric connection, while the remaining PHCs lack even standby facility (generator). Besides, operational telephone and vehicle facility are mandatory at the PHCs to deliver effectual and timely delivery care services to the nearby people. But only

9 % and 12.5 % PHCs of Varanasi district have communication and vehicle facility.

The utilization of health care services has been seen poor in almost all parts of India. Many researchers have pointed out the lack of doctor's residence at PHCs and sub-centres. It is one of the main causes of poor utilization of health care facilties. In the study area, only 40.6 % PHCs are able with doctor's residence. It is also deliberated that a considerable number of PHCs are running without competent doctor. As many PHCs of the Varanasi district are running without the services of doctors (Rai and Nathawat 2013).

Besides, the condition of facilities and equipment's are also not acceptable. There is found substantial variation in these services at development block level. It warrants enhancement in the health care facility (Rai and Nathawat 2013). It is a perquisite for establishing full utilization of health care facilities in the study area.

8.5.3 Distance Constraint

Distance to health care services is a significant aspect for patients, doctors, and administrators. Some of the utmost common functions of geographic information systems are its measurement functions.

The distance is another constraint in the poor utilization of health care facilities in the study area. Out of the total of 1262 villages of Varanasi district, only 37 PHCs are situated in village itself, while 132, 243 and 282 villages lie within 1 km, 1–3 km and 3–5 km distance for the respective PHC's. In general, the majority of the villages, i.e. 568, situated relatively away from PHC's (Table 8.2).

To increase the convenience of these villages, the linking roads should be cemented or metaled (made pucca). Telephone and vehicles obtainable at PHCs must be kept in operational condition, so that the affected people may get the advantage of these services (Rai and Nathawat 2013).

Table 8.2 Distance wise distribution of villages with respect to healthcare facilities

Development block	In village	<1 km	1–3 km	3–5 km	>5 km	Total
Baragaon	7	19	30	28	49	133
Pindra	2	29	35	33	84	183
Cholapur	5	4	10	10	111	140
Chirai Gaon	6	7	20	30	70	133
Harhua	3	38	49	47	32	169
Sewapuri	4	3	17	32	121	177
Araziline	7	28	61	50	63	209
KashiVidhyapith	3	4	21	52	38	118
Total	37	132	243	282	568	1262

Source: Based on Personal Survey

It is detected that about 93.5 % of OPD patients are coming for the primary health centres belonging to the village situated within a radius of 6 km. Out of this, 61.8 % are from 1.6 km. radius. In order to have an impression about the distance wise utilization pattern of health care services, samples have been found for the villages situated within 1 km, 1–3 km, 3–5 km and above 5 km distances (Rai and Nathawat 2013). Table 8.2 obviously represents that the maximum utilization of health care services is originate within 1 km distance from PHC's. As one goes further, the proportion of respondents decreases. So, efforts should be made to decrease the distance between patient and PHC's/CHC's/sub-centres through their suitable and proper place. In addition, ambulance facility should be made obtainable to serious and distant villagers. The road connectivity of villagers must be augmented by connecting them through metalled road (Rai and Nathawat 2013).

8.5.4 Poor Socio-Economic Condition of the Villagers

The socio-economic status of the rural side people is relatively poor than their urban counterpart (Rai and Nathawat 2013). In the study area, the majority of the population belongs to only two religions, i.e. Hindu and Muslim. If one likens these two religious groups, Muslim lie at the lowest rankings of the socio-economic condition, while Hindus are categorized as upper castes, back-ward and SC/ST (Rai and Nathawat 2013). From the castes point of view, higher utilization of health care services is found in back-ward castes followed by SC/ST caste and Muslims. Moreover, they are also economically poor population that is more dependent on PHC's for health services while the well-off people can tolerate high cost for better treatment offered at private clinics, nursing homes and hospitals in urban areas. Hence, they slightly benefited from the services of PHCs. The outcome of education wise utilization pattern of health care services reveals that higher the education, the improved the utilization (Rai and Nathawat 2013). To estimate the affluence in different house types, earning was taken into account in the present study. The well-off respondents use vaccination facility more as compared to economically poor ones. It is stimulating to note that pucca (cemented) and mixed house owners are more sensitive for health of mother as well as their children and size of family. They have been utilizing facilities of mother-child health (MCH) and family planning more than their kachha (unmetaled) house owner counterpart. The occupation wise utilization of health care facilities is very notable. The people who are in job and those who manage agriculture along with jobs visit PHCs/sub-centres rarely (Rai and Nathawat 2013). All these prompts us to improve the services of health care of PHCs/sub-centres, So, sincere efforts are desired to develop the conditions of the villagers through growth of agro-based industries, poultry forming, pig rearing, fish farming etc. In addition, it is required to promote literacy. All facilities should be made available to educate females in the village. It will increase awareness among females to utilize health and family planning facilities provided by PHC's and sub-centres (Rai and Nathawat 2013).

8.5.5 Less Availability of Doctors

The health care facility are delivered by PHCs/sub-centres and CHCs under the leadership of competent doctor. As such, the 24 h accessibility of doctors at PHCs becomes decisive for utilization of all types of health facilities by the villagers (Rai and Nathawat 2013). In Varanasi district it is identified that about 82.56 % PHC's have doctors, the rest depends on other para-medical employees. The causes behind non-availability of doctors at PHCs may be attributed to inattention in posting of doctors at every health care organization and, non-availability of accommodation at these centres. In addition, the non-availability of good convent school for their children in rural areas also comes in the way. Many of the doctors do not join their duty due to these restrictions. Only about 40.6 % PHCs of the study area delivers doctor's residence. As a result of non-availability of accommodation facilities at PHC itself and good schools, majority of the doctors appointed at PHC's exist in cities and visit PHC either for a limited period or on alternate days. The family welfare programme is added very significant service rendered by the primary health centres. This programme wants the service of a female doctor. Even for antenatal checkups, the presence of female doctor is a requirement. But the condition is poor, as female doctors do not wish to help the rurural people because of absence of accommodation facilities and also safety reasons (Rai and Nathawat 2013).

8.5.6 Lack of Awareness About the Services Available of PHCs/CHCs

In rural areas of the district, the level of consciousness is still very deprived. The poor awareness about the health services available at PHCs, sub-centres and CHCs is quite respondents for the lower utilization of health care services delivered by primary health centres (Rai and Nathawat 2013).

In order to have some idea about the consciousness of the presence of various health services at PHC's and sub-centres, 800 people were questioned about the present health services available in the area and their utilization. Only 650

Table 8.3 Awareness about various categories of health services of PHCs/sub-centre

Health services	No. of respondents	Percentage of respondents
Vaccination	489	75.29
MCH	456	70.12
Family planning	405	62.30
Treatment	337	51.92
Disease control	261	40.12

Source: Based on Personal Survey

respondents were conscious of the facilities of PHCs. In the study area about 78 % and 56 % respondents are conscious of the presence of PHC's and sub-centres respectively (Rai and Nathawat 2013). In terms of awareness about several categories of health facilities of PHCs/sub-centres, the maximum consciousness has been noted in favor of vaccination (75.29 %), followed by MCH (70 %) family planning (62 %), treatment (52 %) etc. (Table 8.3).

The above discussion points that the concerted and co-ordinated efforts are desirable to make conscious the people of the study area about the health care facilities obtainable at primary health centres. Besides, satisfactory attention must be paid specially on those people who are either cut off from the print and electronic media or incapable to learn from numerous means of advertisement regarding health services by the state and central government (Rai and Nathawat 2013).

8.5.7 Cause of Dissatisfaction

To augment the rate of utilization of health care services, it is essential to determine the reasons of dissatisfaction (Rai and Nathawat 2013). A total of eight reasons have been shortened according to the view of respondents and the same are given in Table 8.4. Values given in the Table 8.4 show the view of respondents.

Among the various services of PHC's causes, the absence of a female doctor at PHC's is a serious concern. It also ranks on the top of dissatisfaction of the people. Maximum numbers of respondents (354) have opined that poor utilization of health-care services of PHCs is due to the absence of a female doctor. It is followed by distance (352 respondents), non-availability of doctors (288 respondents), absence of diagnostic (Lab testing) services available at PHC's (252 respondents), selective supply of medicine etc. (Rai and Nathawat 2013).

The caste shows a vital role and is also a prime factor in the displeasure of respondents. Upper caste people listed three important reasons for poor utilization of health facilities which include absence of a lady doctor, non- availability of doctor and distance. On the other hand the backward caste people opined that absence of the female doctor; diagnostic services and distance are more accountable for poor state of affairs (Rai and Nathawat 2013).

Contrary to the upper and backward caste respondent's opinion, the outlook of scheduled caste/scheduled tribe go in favor of non-availability of doctor, distance and rough behavior of employees appointed at PHC's. From these remarks it could be concluded that SC/ST people on account of poor economic status don't care much about the health of child and mother (MCH), delivery of child at PHCs etc. Secondly, they are down flattened so they have to assimilate rough behavior of employee. These point gaps are worth noticing for policy creators, planners as well as administrators (Rai and Nathawat 2013).

Table 8.4 Causes of dissatisfaction with the healthcare services of government health centres

Development block	Non availability of doctors	Absence of lady doctors	Time limit	Distance	Absence of diagnostic facilities	Selective distribution of medicine	Rough behavior of staff	Demand of money
Baragon	37	45	31	29	35	19	17	13
Pindra	33	53	37	25	31	15	11	11
Cholapur	37	59	43	33	11	11	13	17
Chiraigaon	41	49	33	37	43	17	19	13
Harhua	35	51	33	49	39	19	17	09
Sewapuri	39	29	31	59	41	21	19	17
Araziline	29	37	29	57	17	23	15	19
Kasi Vidyapith	37	31	33	63	35	17	17	15
Total	288	354	270	352	252	142	128	114

Source: Based on Personal Survey

8.6 Education and Training

There is necessity for formal education and training at all levels for appropriate utilization of health services among the villagers. Education of public as to what constitutes good nutrition. It should be based on what is virtuous and obtainable (Rai and Nathawat 2013).

(1) Most of the people in the study area are superstitious as such they should be made familiar with the modern scientific methods and approaches of medical treatment. In order to attain the objective importance should be laid on health education and public inspiration. In this direction mass media like radio, television, newspaper, magazine etc. should be come forward in making consciousness among people towards mother and child health care and the present health services in the Varanasi district (Rai and Nathawat 2013).
No voluntary group has come forward to instruct consciousness among villagers towards health care, nor any effort has been made to influence people for group action. Most of the programmes started by the Government lack the basis factor of decision making by the local people. Therefore, efforts should be made for public participation nurse, extension educator and even doctor should come forward to create consciousness about health and sanitation, illness control programmes (Rai and Nathawat 2013).
(2) There should be proper removal of polluted water and solid waste. It will help in reducing the incidence of illness (Rai and Nathawat 2013).
(3) There should be transparency regarding the medical facilities, supply of medicine to all sorts of people. Care should be paid in providing satisfactory services and specialized facilities for medical treatment in Government hospitals (Rai and Nathawat 2013).

The health workers should be asked to visit his area at least once in month. He should be proficient to take weight of newly born child. The difficulties of health care are vast (Rai and Nathawat 2013). Access to primary health care is insufficient to the majority of the people because of low obtainability of basic preventive and promotive health care packages, clinics, doctors, drugs and para-medical personnel in rural part of the district. Greater stress on preventive health care, medical and health education should be laid. Health literacy efforts should be made integral to preventive, promotive, curative and rehabilitative health care. A meaningful participation of the private sector and NGO's in critical in all these actions for promotive a people-oriented and a sustainable health care system (Srinivasan 1984; Rai and Nathawat 2013).

Organization of health services has become complex, centralized and insensitive to varying health felt-needs of the rural community. It is recommended that organizational set-up of health facilities needs reorganization (Rai and Nathawat 2013). While the health organization has grown greatly in the past five decades, functionally the structure has not changed with the active and different demands of effective is weak because of the low status accorded to training in public health, epidemiology and health management and insufficient decentralization of consultant and resource distribution. The utmost significant problem is circulation of the health manpower/health human resource, both geographically and category-wise. Both technical knowledge and skills and motivation to help the rural people fall short of prerequisite and expectations (Rai and Nathawat 2013).

Impartial distribution of rural health care facilities for ensuring equity for health care should be ensured by the government. Location of health facilities and services should be such that these are easily available and obtainable to the rural community (Government of India 2001, 2006, 2009; Rai and Nathawat 2013). In view of the experiences and problems faced in the delivery of rural health facilities, it has been understood that acceleration of the pace of the execution of rural health programmes in a vital imperative and intensive efforts necessity to be made by the State and Central Government. Additional, substantially large financial investments needs to be made for speedily improving the health profile of the country. For making rural health care facilities more significant to the rural community, it is required to bring about fundamental changes in the focus and complete health care delivery system in general and rural health care, in particular (Goel 1984; Planning Commission 2001; Rai and Nathawat 2013).

The researchers hopes positive provided the aforesaid limitation are minimized and recommended remedial measures are incorporated whenever/whatever feasible.

8.7 Action Points

- It is essential to enhance the health care services and transportation network in the study area.
- To sustain the zero death due to vector born disease e.g. malaria and other diseases.
- Early detection and quick response measure though more rapid surveillance and timely delivery of medicines & other supplies in peripheral health institution.
- Inter sectoral coordination particularly with the Department of Local Bodies to ensure removal and control of breeding of the vector of malaria or other vector borne diseases in both urban and rural areas to minimize the problem of vector born disease throughout Varanasi district.

There is clear indication that the application of geographic information systems (GIS) for health mapping can play a key role in public health management. Mapping tools can be used to support decision-making and planning in epidemiological studies, disease surveillance, disaster management, health services, environmental health assessment and social determinants of health.

It is essential to develop an institutional framework, policies and procedures compulsory to support regular health data collection, health mapping and reporting to inspire evidence-based policy-making and planning at all levels and establish health mapping units with the necessary infrastructure and resources, including human resources, to support health mapping activities at state level and collaborative efforts at district and development block level.

And also develop integrated national systems for the management of health data and link the systems to digital maps, allowing technical programmes to input, verify, update and view health data on maps online.

References

Datta, S. P., & Kale, R. V. (1969). An operation research study in primary medical care in Pondicherry. *Indian Journal of Preventive Social Medicine, 1*, 65–72.

Dutt, P. R. (1962). *Rural health services in India. Primary health centre*. New Delhi: Central Health and Family Planning.

Gatrell, A. C. (2002). *Geographies of health: An introduction*. Oxford: Blackwell Publishers.

Gatrell, A., & Loytonen, M. (1998). *GIS and health*. London: Taylor & Francis.

Goel, S. L. (1984). *Report of planning commission, public health, administration*. New Delhi: Sterling Publishers.

Government of India. (2001). *Rural health statistics*.

Government of India. (2006). *Rural health statistics*.

Government of India. (2009). *Rural health statistics*.

Kumra, V. K., & Singh, J. (1994). Assessment of community attitude regarding the services of primary health centre: A medical geography study. *Indian Journal of Preventive Social Medicine, 25*(3/4), 118–126.

Planning Commission. (2001). *Draft mid-term appraisal of the ninth five year plan, 1996–2001*. New Delhi: GOI.

Prakasam, C. P. (1995). Utilization of primary health care services in a rural setup. *Health Population Perspective Issues, 18*(2), 72–80.

Rai P. K., & Nathawat M. S. (2013). GIS in health care planning: A case study from India. *Forum Geographic Journal (Romania), 12*(1), 152–163. Available at: http://dx.doi.org/10.5775/fg.2067-4635.2013.180.d

Rai, P. K., Nathawat, M. S., & Singh, S. B. (2011). Role of GIS in analyzing the distribution of health care services. *National Geographical Society of India, 57*(3), 15–24. (ISSN: 0027-9374).

Rai, P. K., Nathawat, M. S., & Onagh, M. (2012). Application of multiple linear regression model through GIS and remote sensing for malaria mapping in Varanasi district, INDIA. *Health Science Journal (International Journal of Nursing Research and Review), 6*(4), 731–749. (ISSN 1108-7366E). Available at http: www.hsj.gr/volume6/issue4/6414

Richards, T. B., Croner, M. C., Rushton, G., Brown, C. K., & Fowler, L. (1999). Geographic information system in public health. *Public Health Reports, 114*, 359–373.

Rohan, F. (2002). Assessing the applicability of GIS in a health and social care setting: Planning services for informal carers in east Sussex, England. *Social Science & Medicine, 55*, 79–96.

Sinha, S., & Rajeswari, S. (1993). Social structure and utilization of health care facilities in rural Haryana. *Annal NAGI, 30*, 37–47.

Srinivasan, S. (1984). *A primary health centre in Tamil Nadu*. Kurukshetra, p. 15.

Kolibri 360 GmbH
Ferdinand-Jühlke-Straße 7
99095 Erfurt, Deutschland
produktsicherheit@kolibri360.de